RESTAURANT SERVICE:
BEYOND THE BASICS

Wiley Professional Restaurateur Guides

RESTAURANT SERVICE: BEYOND THE BASICS

Carol A. Litrides   ❖   Bruce H. Axler

John Wiley & Sons, Inc.

New York • Chichester • Brisbane • Toronto • Singapore

To Mom and Dad for their
unending support, to Bruce, my co-author, for his help and encouragement,
and to Kim and Cathy for their days and nights
of typing, typing, typing . . .
Thanks!

C.A.L.

Preface

You own, operate, or are employed at a stand-alone restaurant, or at one within a hotel, theme park, area attraction, or other hospitality establishment. And there's another restaurant next door and one across the street. You serve fresh food in a friendly manner and in a convivial atmosphere. So do they.

♦ What can give your establishment a unique identity?
♦ What will make guests return to your establishment?
♦ What are the "difference makers"?

This book answers these questions and others. We look at a guest's first impressions of a restaurant and the employees who can affect those impressions—hosts,* reservations agents, coat-check personnel.

We examine special services restaurant managers must be familiar with, such as lost-and-found, the implications of the Americans with Disabilities Act, the positive impact of employing bilingual employees.

After your guests are seated, what can make a difference in their dining experience?

We suggest selling and marketing approaches to specific customer market segments, including those who do not wish to drink alcoholic beverages, those who desire exceptional, even rare, wines, and those who enjoy ceremony with their beverage service, for example, "high tea" or "flaming coffee drinks."

Our first book *Food and Beverage Service,* covered the basics; this book goes beyond the fundamentals of food and beverage service to explore the "difference makers." Today's successful establishment not only offers great service in every way expected but

---

*Throughout this book, the term *host* will indicate employee greeters (hosts and hostesses) of either sex. This use is for purposes of simplicity only, and readers may freely incorporate the term *hostess* if desired.

makes sure it gives guests a reason for going out of their way to make a trip there again. The successful business goes beyond the basics to provide that extra "sizzle" guests search for, that ephemeral "something" guests remember and will return for.

That's the focus of this book. We describe what we mean by "beyond the basics" of good service and suggest ideas to try in your establishment so guests will go out of their way to return there. To expand the book's usefulness, we also explain what to look for when you are a guest at hospitality establishments.

We sincerely hope you enjoy it.

CAROL A. LITRIDES
BRUCE H. AXLER

New York, New York
February 1994

Acknowledgments

Our utmost thanks to the many people who have assisted in providing materials for our use in this book. Special thanks to those who have donated their time and efforts to make this book relevant and interesting:

Barry Goodman, who has offered his rare wines for tastings and the use of his rare wine catalogs; Larry Leckart for his wonderful marketing stories and caviar service assistance; Mary Obarowski and Ricardo Oliver for their assistance with the Accessibility Checklist in Chapter Four; Lynn Milet for her technical support; Brian Fitzpatrick, of Fitzpatrick Winery, for his input regarding organic wines and organic farming; Jim Litrides for his assistance on the Safe Water Drinking and EPA water standards; and Mom Litrides for the Turkish coffee recipe . . . which we've enjoyed many times.

In addition, we'd like to thank the following:

Ron Dain, Vice President, VanSan Corporation, for allowing us to reference its *Omni* model lectern, which is able to adjust to several heights, in our ADA section.

Arno Ford Furstenberg, Director of Design, CAS The Sign Company, who pored through the *Federal Register* to find information regarding ADA (Americans with Disabilities Act) signage verbiage, and who provided the signage artwork for this book (see Chapter Four and Appendix A).

William Leigon, Vice President, Ariel Vineyards, for input regarding dealcoholized wines and the listing of Awards (see Chapter Eight).

Serena Sutcliffe M.W., Head of Wine Department, Sotheby's Auction House, London, for the use of the ullage figure and information from their Wednesday 18th November 1992 catalog (see Chapter Eleven and Appendix B).

J. Michael Broadbent, M.W. Wine Director, Christie, Manson & Woods, Ltd., London, for permission to use the ullage figure from the Thursday 9 April 1992 *Fine Wines and Vintage Port Catalog* and verbiage throughout Chapter Eleven and in Appendix B.

Tom Baines, Director of Marketing, Novon Company, Inc., for permitting reference to their Dynamic Blending Mixer® with photograph (see Chapter Eight).

Contents

xi

PART IV   A PRIMER ON
❖   WINE SERVICE

❖   PART V   THE STAFF

P·A·R·T I

GUEST COURTESIES: THE DIFFERENCE MAKERS

O n e

First Impressions

Hosts (and/or hostesses) are often the first people guests meet in an establishment. They create the initial impression of that establishment, representing the owner, management, and staff to the public. They also are the last people guests see and hear when leaving the establishment—creating a lasting impression—good or bad. It is obvious why this position is as important to the success of any establishment as the food, the ambiance, and the service!

A restaurant guest's first impression should begin with the greeting offered by a smiling, friendly person. It is important for the host to be well groomed and to avoid using discernible perfumes or other products with strong odors that can conflict with the wonderful smells of the fresh flower arrangements and the food itself.

Besides offering a greeting, the host should be able to assist with information about the establishment as well as about other businesses, sights, or historical items in the area. Hosts as well as others who deal directly with guests should be ready to provide the following kinds of assistance:

♦ Representing the establishment to the public, in appearance, demeanor, and actions.
♦ "Selling" guests on the establishment, so they want to remain, then return.
♦ Directing guest traffic.
♦ Confirming and making reservations.
♦ Assisting the maître d'hôtel, if there is one.
♦ Answering inquiries about the restaurant, property, and area in general.
♦ Answering inquiries about the menus, meal periods, current and future reservations.
♦ Encouraging use of gift shop or cocktail lounge facilities by guests who are waiting for a table.

3

MEETING AND GREETING GUESTS

❖

Because every guest who enters the establishment should be greeted by a smiling host and/or maître d'hôtel, the owners and/or managers should agree on suitable wording for the greeting. Often, a simple "Good afternoon" or "Good evening," or other similar acknowledgment of the guest's presence is sufficient. If the establishment has a casual atmosphere, something more informal might be appropriate (e.g., "Howdy, folks" for a down-home atmosphere).

If reservations are suggested or required, guests should be asked whether they have a reservation. After an affirmative reply, the next question is, "Under what name?" The exact time of reservation, number of persons in the party, and the requested dining room (if there is more than one) should be ascertained from the reservation sheet.

If the restaurant uses seating cards for diners in the à la carte areas, a card should be completed with the following information:

♦ Dining/banquet room.
♦ Reservation name and phone number.
♦ Number of guests in the party.
♦ Time of reservation.
♦ VIP (Very Important Person/Party) notation if appropriate.
♦ Special considerations (e.g., birthday cake or wheelchair access required).

At a minimum, the card should list:

♦ Name of party (center of card).
♦ Number of guests in party.
♦ Time of reservation.

Guests should be directed or escorted by a host or maître d' to the appropriate dining or banquet area. Standardized directions should be prepared for guiding guests—every employee should use the same verbiage when giving directions. Print and post the directions to each dining space, and give each host and maître d' a copy. Ask the employees to practice giving these directions to each other until the instructions are memorized and become automatic.

Guests who do not have reservations in the à la carte dining rooms should be asked whether they are attending a private party. Where there are several private dining rooms or banquet halls, it is necessary to follow up on an affirmative answer by asking, "Which party?" Directions to the proper room should then be given, including which coatroom to use.

Walk-In Sheets

Guests without reservations should be advised of the approximate wait time for a table. The host should register the name of the party, the number of guests in that party, and the time of arrival on a "walk-in" sheet, and the guest(s) should be directed to the coatroom, then to the gift shop and/or cocktail lounge. Often a host handles these activities under the direction of the maître d' or a "lead" host.

The walk-in sheet should have several columns, each of which designates a table size (known as a "top") in the establishment. There should be one column for two-tops, or "deuces" (tables that seat two guests), one for four-tops, one for six/eight-tops, and one for large parties. As a couple without reservations enters, their name and time of arrival are placed under the "deuce" column; if a party of eight without a reservation enters next, they become the first entry in the "six/eight-top" column, and so on.

The deuce "wait line" will move the fastest. This is because tables for two tend to turn over more quickly than other tables, there are often more two-tops and four-tops than any other type of table, and couples can be seated at a deuce or at a four-top. As a general rule, if there is an empty four-top and no reservations for a four-top are expected within the next half-hour, that table can be given to the next waiting deuce reservation. Likewise, a six- or eight-top can be given to a waiting group of four or six. It is inadvisable, however, to squeeze a group of six around a four-top table.

At a full-service, leisurely dining restaurant, a deuce usually turns over in about $1^1/_2$ hours; a four-top in 2 hours; and a group of six or more in about $2^1/_2$ hours. For "special occasion" tables (e.g., a birthday luncheon or dinner, a bridal or baby shower luncheon with presents to open or photo albums to be passed among the guests), 3 hours might be blocked out. To gauge the seating capacity and plan seating charts, each establishment will want to track average "turn times" for various table types.

Walk-In Seating Cards

If seating cards are used in addition to the walk-in sheet, a "walk-in" seating card with the usual information (name, time of arrival, number in party) should be prepared. The seating cards should be placed in chronological order, and grouped by party size and type of table—deuces, four-tops, six/eight-tops, and large parties.

As appropriate tables become available, the walk-in parties should be seated accordingly. (A group of six people might have arrived at 6:00 P.M. and a deuce at 6:10 P.M., but if a deuce opens up first, the couple will be seated first, even though they arrived after the group of six.)

Walk-in seating cards are usually plainer than reservations seating cards so as to differentiate them. For instance, an establishment with four dining areas might have five seating cards: red-bordered cards for the Red Dining Room; blue-bordered cards for the Blue Dining Room, and so on; the walk-in cards, however, would be plain—unbordered. Often, a logo will be imprinted on the reservations seating cards, or perhaps a different logo for each dining room.

A party's VIP status is commonly designated on the seating card by affixing a gold star or a "VIP" notation in one corner. This status should be noted in some manner so the host, maître d', captain, and all members of the service team can quickly identify such parties. Since the person who checks guests in on the reservations sheet may not be the same person who escorts the party to their dining room and seats them, the VIP designation on the seating card assists greatly. As the guests are seated, the maître d', hostess, or captain—whoever is seating the guests—should write the party's table number on the seating card and return it to the sidestand or host station.

Using plain cards for walk-in parties alerts the maître d' or host that the group can be seated only if a table is available, that is, if a reservation party doesn't show or there are "extra" or unreserved tables. If upcoming reservations are not carefully accounted for, walk-in guests might be offered a table needed shortly thereafter for a party with reservations.

Usually, reserved tables are held for 15 minutes before being given to the next waiting walk-in party. Whatever the "hold" time, be sure guests making reservations are aware of that limitation and realize that their table will be given to the next available group of guests when the time is up if they have not yet arrived. While walk-in guests are waiting for a table, the host can direct them to the cocktail lounge and/or gift shop.

Hosts should know the locations of all restrooms, telephones, water fountains, and coatrooms, and how to give directions to each place. In addition, they should be able to direct guests to the business office, the lost-and-found, the banquet office, and the employment office, if there is one. If some restrooms have baby-changing areas and/or facilities for the physically disabled, the host should also be aware of these facilities and able to tell guests where they will find these services.

Banquet and Private Dining Facilities

Restaurants may have banquet and private dining rooms. Sometimes, these private dining areas may even have their own restrooms, public telephones, and coatrooms nearby. Perhaps the guest of honor or an invited guest is a celebrity—a political figure, sports hero, or entertainer desiring privacy and a dinner (or lunch) away from photographers, the press, and/or fans. For these and other reasons, the restrooms, coatrooms and public telephones for such dining rooms may be reserved for only those guests attending private functions. If this is the case, it is important to ensure that the general public does not frequent those facilities, but utilizes alternate locations designated for public use. Where certain restrooms and pay phones are reserved for private parties held in the banquet rooms, inform the staff so they will direct only guests attending such functions to those facilities.

❖ SEATING GUESTS

Table Preparation

When a table becomes available and has been fully reset, the host should take the seating card of the guests assigned to that table and lead them to it. Management should devise a signal whereby the service staff can let the host know when a specific table is ready for guests. Some establishments use table napkins—if the napkins are in place, the table is ready; if not, setup is still incomplete.

A table that looks ready for guests from a distance may not be; perhaps certain silverware pieces, such as forks, are missing, and the service team is obtaining more utensils from the warewashing station. Thus, waiting for the agreed-on signal may avoid delays and customer irritation once guests have been seated.

The maître d' of the dining room where the guests will be seated often makes the table assignments.

Host's Procedure for Seating Guests

Locating the Guest Party

With the seating card in hand, the host should locate the party by paging them. Use a loud, but not shouting voice, and call the name of the party (e.g., "Blackstone party"). If you locate them in the lobby or gift shop, escort them to their table and seat them. Write the table number on the seating card and return the completed seating card to the sidestand or host station.

For guests who are in the cocktail lounge or outdoor cocktail area, you need to know whether bar tabs can be transferred to the dinner check or must be settled prior to the guests' being seated. If the bar tab can be transferred to the dinner check, simply escort the guests to their table. Then return to the cocktail lounge and transfer the bar tab onto the dinner check if that duty is assigned to the host. If it is the bar server's responsibility, give the server the table number where that party has been seated so the bar tab can be added to their check.

If the bar tab must be settled before seating guests in the dining room, explain this policy to your party by saying, "Your table is [shortly will be] ready. Would you please settle your bar check, and I'll be back in a moment to seat you?" Leave the party in the cocktail lounge, and give them a few minutes to pay the check.

Many establishments will allow most of the party to be escorted to the table, leaving one person in the cocktail lounge to pay the bar tab. You should then check back for that individual, and seat the last guest with the rest of the party.

Beverages may need to be physically transferred from the cocktail lounge to the dining room, and it is preferable to use a standard procedure for this task. When you are not sure of drinks' names and appearances, and aren't sure what drink is in which glass, there are clues to remember "who's drinking what" so you won't have to ask the guests to identify their own drink when they reach the dining table. You might say to yourself, "The woman in

---

PROCEDURE FOR TRANSFERRING BEVERAGES

1. Find a beverage tray.
2. Hold the tray on the flat palm of your left hand.
3. Transfer the drinks from the guests' table onto the bar tray, keeping in mind who's drinking what drink (if it's obvious).
4. Lift all glassware by touching only the stems of the glasses (if stemware), or else by touching the bottom one-third of the glasses; do not touch the upper portion of any of the glasses.
5. Proceed to the table, asking the party to follow you (after ensuring that the bar check has been paid if that is house policy). Wait for the guests to seat themselves, then place each person's drink on the right side of that guest, using your right hand to place each drink. Walk in a clockwise direction around the table if a round or square table and it is possible to do so.

pink is drinking from the half-filled highball glass; the woman in blue is drinking from the champagne glass; the man wearing glasses has the rocks glass with a twist; the man in the polo shirt has a plain rocks glass." You might even write on a small pad:

Pink W: highball	Blue W: champ
Glasses: rocks, Tw	Polo: rocks

Don't rely on placing the drinks on your beverage tray in the same order that the guests were seated in the lounge—they might seat themselves in a different order in the dining room.

If you are not transporting beverages, assist in the seating of the guests by pulling out the chair of at least one person. Keep in mind the general rules of seating: seat women, the elderly, and children before able-bodied men:

Ladies before gentlemen;
Old before young.

Using Seating Cards

Once the guests are seated, write the table number on the seating card and place the card on the station sidestand. This enables the captain of that station to know the following:

♦ Name of the party.
♦ Number of guests in the party. (If a place setting must be added or removed, it can be done without asking, "How many will there be in the party?")
♦ Table number where they are seated.

If the table number has been changed from that initially assigned to the party (e.g., the guests requested a different table after being led to the table by the kitchen entrance), note the new table number on the seating card. Otherwise, the captain is likely to approach the wrong table, address the party by the wrong name, and look for the wrong number of guests.

Seating cards should be kept for reference (but out of sight) in the event that someone in the party receives a telephone call while dining. If the caller knows the name of the party, as shown on the reservation sheet, the manager or maître d' can, by reviewing the reservation cards, determine where that group is seated and bring the message to the guest.

As soon as the party departs, the seating card is disposed of to avoid confusion with later parties seated at the reset table.

PREFERRED TECHNIQUE FOR SEATING GUESTS

1. Pull out the best seat, for example, the seat facing a window with a view.
2. If there is a woman in the party, offer it to her; or to the eldest woman of a party if two or more women are present.
3. Assist the women with their seats if the men in the party do not assist them.
4. At wall tables, pull the table away from the banquette (sofa seat), so the women in the party can gracefully seat themselves. Return the table parallel to the wall. Seat the men in the chairs along the aisle, facing the women.
5. If there are not enough chairs for the party, bring the nearest unoccupied chair to the table for the remaining guest(s). After seating your party, bring replacements to the table from which you borrowed chairs so it can be fully seated. If you bring spare chairs from a storage area, or move them from along a wall, return them to their original location after the party has departed.
6. Mentally note your guests' table number and verify that the same number appears on the seating card. If not, change it, so the card is correct.

❖   # GIVEAWAYS POLICY

Balloons, lollipops, roses on Mother's Day, flags on the Fourth of July—all these are typical giveaways to restaurant guests. Employees must know what's available, to whom, and when. What are the limitations, if any? Can the guests purchase additional giveaways if they desire more than one item?

For instance, a balloon policy might be as follows:

Balloons are available during lunch and dinner meal periods only. They should be given without request (automatically) to children, and to persons celebrating a special occasion (e.g., birthdays). They can generally be given to others, but only on request.

If children are present, the server or busser presents one balloon (or other giveaway) to each child. Usually, the policy is to provide one balloon for every two adults in a group celebrating a special occasion although some establishments give a balloon to each person.

Since adults often leave balloons or other giveaways at the table, using one balloon per two adults avoids waste and reduces cleanup time.

Should someone in the party request more balloons or other giveaways, bring more, without charge, up to a maximum of one per person. Additional balloons or other giveaways should be available for a modest charge to cover the cost of the item and discourage excess waste.

Guidelines for quantities and methods of distribution need to be devised for all giveaways. For example, management can have servers bring lollipops to the table with the check, or can locate the candies at the cashier stand where children can help themselves as they leave.

If one rose per woman is the giveaway for Mother's Day, are the roses distributed at the door when guests enter or presented at the table? If at the table, does a "character" (dressed perhaps in tuxedo and top hat) hand a rose to each woman, or does the server offer the flower when taking the order? If a chocolate Easter egg is the giveaway, does an employee "Easter Bunny" give out the treats at the door, or is there a search-and-find display that children can rummage through to choose an egg?

Giveaways often become seasonal traditions and they attract repeat customers. If you always have a "character" employee, (whether it's Santa Claus, an elf, the Easter Bunny, Uncle Sam, a turkey, a cupid)—if you have a character for children, moms, dads, and other guests to talk to and even pose for photos with—you have an attraction, a draw. You're going beyond the basics!

❖ PROVIDING INFORMATION AT THE FRONT DESK

It is vital that the front desk or "greeting area" personnel be able to provide the following information:

- ♦ Neighborhood attractions, historical locations, and so on, with directions to them.
- ♦ History of the location, building, owner(s).
- ♦ Meeting room areas.
- ♦ Dining areas: À la carte rooms, banquet rooms, outside gardens, counter, and takeout areas.
- ♦ Floor plans of all rooms; table numbers within rooms.
- ♦ Type of cuisine served and average check in each dining area.
- ♦ Dress requirements for all areas.

- Minimum charges, cover charges, corkage fees.
- Seating policy.
- Days and hours of operation.
- Holiday service, if any.
- Reservations procedures.
- Types of service: Banquet service, à la carte, group sales, counter, takeout.
- Gift certificates.
- Deposits, prepayments.
- Procedures for handling job applicants.
- Lost-and-found procedures.
- Management names, titles, and phone extensions.
- Management of parking lot: Concession or establishment; free, valet, or stamped ticket.
- Emergency numbers (police, fire, etc.).
- Location of fire extinguishers, all exit doors, and all telephones.
- Front-of-house functional areas.

The following sections describe some types of information that every host and front desk employee, reservations agent, and manager must know and should share with other employees.

Interesting Historical Trivia

Every employee should be familiar with the history of the establishment. The recital of this history should be brief and entertaining, with information to stir guests' imagination. Many tourist spots have mementos from famous personalities who have frequented the restaurant, or can provide a colorful story about the founder of the place. If the restaurant was built on an unusual site or has evolved from a former use (e.g., factory or train station), that fact might spark interest.

Some establishments have invented "histories" just to create a "point of difference" or conversation piece for customers. The management can display this folklore on a conspicuous poster in the front lobby, or have it printed on the front inside cover or the back of a menu, on a table tent, or on a placemat for guests to read while awaiting their food or beverages. This "history" can become a great word-of-mouth advertisement.

Dining Areas and Floor Plans

Every host, maître d', and server should know each room's special characteristics and floor plan as well as its seating capacity. Special decorations, wall treatments, window treatments, and interesting historical details should all be noted. For instance, there might be a mighty oak tree within view, and the server could say, "Local tradition says that Daniel Boone planted the acorn for that tree when he was exploring the Wilderness Road." Indoor as well as outdoor dining areas should be considered as possible points of interest.

Floor plans designating areas of the facility used by employees and the public should be available as well as larger, individual floor plans of each dining area and the kitchen, designating food and beverage pickup stations, espresso stations, and so on. In the individual dining room floor plans, server stations and placement of tables, chairs, and sidestands should be designated.

Cuisine

Types of cuisine (Italian, French, American, Mexican, etc.) must be known, as well as whether the cuisine is gourmet, "health-food," vegetarian, or a combination. Any special menu items, such as low cholesterol, low-fat, or specialty items, should be communicated to the staff.

Restrictions: Dress Requirements, Minimum Charges and Cover Charges, Seating Policy

Dress requirements may mean jacket and tie are obligatory (more formal restaurants) or simply that shirt and shoes must be worn (seashore resorts, casual restaurants). These regulations should be prominently posted for guests.

Customers need to be aware of minimum charges, if any, at the table as well as cover charges or entertainment charges, if these are applicable. A written statement should be displayed at each table (e.g., on a table tent, each placemat, or the menu). In addition, the server, on first approaching the table, should reiterate these standard charges.

The management must specify whether certain areas are reserved for diners having a full meal and other areas are reserved for patrons having light fare or only cocktails, or desserts and coffee.

Many operations, during peak meal periods, restrict dining room seating or garden seating to those having full meals but allow guests having only cocktails or coffee and dessert to be seated in those sections if tables are open during nonpeak times. Other operations have firm policies and strictly segregated areas: the cocktail area, the dining room, and a dessert parlor or light dining area.

Usually, those with reservations have first priority for a table; walk-ins waiting for full meals have second priority; last to be seated are walk-ins desiring light refreshments.

Dealing with Questions of Time

Days and Hours of Operation

Every employee needs to know how many days a week the operation is open and whether it is open on holidays. What meals are served—breakfast, brunch, lunch, dinner, supper? During what hours and for how long does the restaurant serve each kind of meal?

Reservations

Are reservations suggested, necessary, or not accepted? If accepted, when is the reservations office open (what days, and what hours per day)? What is the telephone number?

Where there is no reservations office, the front desk, a host, or the manager often takes responsibility for reservations.

How far in advance are reservations accepted? Anywhere from 2 weeks to a month is usual for à la carte dining rooms; up to 2 years in advance may be suggested for meeting and banquet spaces. Are there special procedures for holidays such as Mother's Day, Thanksgiving Day, Christmas Day, New Year's Eve?

What time is the last seating? When does the kitchen close? Is it closed between lunch and dinner, or is there continuous service? If the restaurant closes at 10:00 P.M., is the kitchen open until 11:00 P.M. for guests who enter at 9:55 P.M., or may guests only order hot food until 9:30 P.M. or whenever?

Management and/or the owners must determine these policies before opening for business to avoid creating guest expectations that cannot be met. Consider the ramifications if guests, who have been told earlier in the day that the restaurant is open until 10:00 P.M., arrive at 9:55 P.M. expecting to order a full dinner and are told, "The restaurant is only serving dessert and coffee at this time." Would you return?

Are guests allowed "grace minutes" on a reservation before the table is released to seat another party? If so, how many minutes?

Are there any time limitations for accommodating walk-ins (e.g., only until midnight)?

Who handles the person who visits the restaurant to make reservations for another date or another meal period? Where there is a house phone and a separate reservations department, it is fairly simple to tell the guest, "I'll connect you directly with the reservations department, and you can make your reservation. One moment please." After getting through to that department, explain to the reservationist that a customer is standing by to make a reservation; then put the guest on the line.

Types of Service

Private Parties

Is space available for banquets (e.g., weddings, anniversary parties)? If so, who should be contacted? Often there is a banquet director or banquet sales representative, and the guest can be directed to the proper office, transferred to the office by telephone, or given the representative's business card with the phone number for inquiries. Before escorting a prospective client on a tour of available banquet spaces, check the banquet sheet to ensure you don't interrupt a board meeting or other private party.

On such tours, you should review each of the banquet rooms, and its decor or style, with the potential customer. If a date for the event is known, stress the advantages of the rooms available for that date. Some guests have a favorite room and might want only that accommodation for their event. This is especially true for weddings. If a guest is inflexible about the room choice and it has already been booked, it is worthwhile to suggest changing the date of the planned celebration. Otherwise, an alternate room is the only option unless the party that has already booked the room is willing to change locations.

Sometimes a party who wants a particular room is willing to pay a premium for it. The establishment can then ask the party with the earlier reservation whether another room at a reduced charge would be acceptable (the normal charge less the premium the other party is willing to pay for the room of choice). If the room choice is not critical to the people who first booked the room, they are often willing to change locations for a discounted rate.

Other establishments reserve the right to assign rooms for scheduled banquets or private parties depending on availability and appropriateness of the facilities. In such cases, a potential customer who deems a specific room to be essential for a special event can insist that the use of those premises be guaranteed in a written contract.

Types of banquet service available and the pricing of each type of service should be reviewed—sit-down service, a buffet, a cocktail reception, a dinner dance, or a combination. Perhaps a business meeting is being scheduled in conjunction with a meal. A typical arrangement is a business breakfast (usually buffet) followed by a kickoff meeting to describe new products or services. The presentation may require the use of audiovisual (A/V) equipment, and demonstration products may be scattered around the room for guests to try, look at, and sample. The company might finish the meeting with either a sit-down or buffet luncheon.

When A/V equipment is needed, the restaurant often supplies it for the customer by subcontracting with a company specializing in such products. The A/V company will provide equipment to the party's specifications, deliver and set it up, test it, then tear it down, pack it, and remove it when the meeting is finished. The A/V company charges the establishment, the establishment usually adds a surcharge, then adds the total A/V charge to the party's invoice for the event.

If a wedding is planned, a bridal room should be available for the bride and bridal party to change clothing and freshen up before walking down the aisle. Another important issue is, who will take responsibility for gifts brought to the reception? Will there be a separate table for them? Can they be locked in the office overnight and picked up the next day, or must they be removed immediately after the event?

What special menus are available? For example, are there menus for children's parties including hamburgers, hot dogs, peanut butter-and-jelly sandwiches, individual pizzas?

The restaurant may receive requests to arrange diverse services for guests, such as flowers, musicians, magicians, clowns, and/or photographers. Every time the establishment successfully handles such services, it is going beyond the basics. It is dealing in difference makers.

Group Sales

Group sales, if available, are usually sales to groups of more than 15 people, often on a group tour, who want to dine at the same time. Often there is an event before or after the meal, e.g., going sightseeing or shopping. (Or, they've just come from sightseeing or shopping, are taking a meal break, and then will resume sightseeing or whatever.)

Group sales usually involve having preset menus that the kitchen can prepare ahead in large quantities. There are usually two options: (1) Each member of the group chooses among three or four entrées or (2) all members of the group have the same entrée (with the exception of vegetarian platters, which are prepared on request if the kitchen has sufficient notice).

Usually the appetizer and the dessert are the same for the whole group. The easiest for the kitchen, and thus usually the best buy, is a predetermined menu, with no individual choices—everyone gets the same appetizer, entrée, and dessert. Doing this, the entire party might be able to order prime rib roast dinner for the same price that a chicken or stuffed eggplant dinner would cost if the group ordered from a selection of three or four entrées.

Regardless, group sales involve a limited menu selection and a specified time period (usually 2 or 2½ hours from start to finish). If the party goes over the allotted time, there is usually an additional charge for cutting into the turnover time needed to break down and set up the room for the next function. It is important to agree on this stipulation and write it into the contract before the guests arrive and the function begins.

Busy establishments often allow only an hour for breaking down and resetting for the next function. A group that goes over its time frame complicates things for the next party. Even if the tour bus arrived late, the passengers usually must be out of the dining room by the stated time. For instance, if a party had booked lunch from 11:30 A.M. to 1:30 P.M., but the bus didn't arrive until noon and the guests weren't seated until 12:15 P.M., they would still need to depart by 1:30 P.M. or be assessed an additional charge. The manager sometimes waives the additional charge, however, if another group is not booked immediately following the "late" one and house policy permits such leniency.

There is usually a group sales manager who books parties, reviews menus with tour group leaders, oversees the actual service of the groups, and handles the billing for such sales. Thus, anyone inquiring about a group sale is referred to the group sales manager, who arranges a meeting either in person or by telephone.

Gift Certificates

Common questions about gift certificates include the following:

♦ Are gift certificates available?
♦ In specific denominations or can the customer choose?
♦ How long are they valid from the issue date?
♦ If the total amount is not used, what is the refund procedure? (Usually a check is mailed to the person whose name is on the certificate if the amount of the refund is one dollar or greater.)

The front desk should be able to deal efficiently and pleasantly with all these questions.

Deposits and Prepayments

A routine should be set up for handling prepayments and deposits. Often, prepayments involve a parent prepaying for a son or daughter's graduation dinner or birthday party with a few friends. A similar situation would be a business person prepaying a business meal so no check is presented at the table.

Handling Job Applicants

There should be an established procedure for responding to people who seek employment. Application forms should be kept in an accessible place. Rather than having the host state, "We are an equal opportunity employer, and we consider all qualified candidates for any positions for which we are hiring" this disclaimer should always be in writing on the application form.

Posting current job openings can eliminate many inquiries. Job openings usually are posted for internal applicants first, and after a period ranging from a week to a month, they are offered to the general public. Most questions should be directed to human resources department personnel or the manager on duty.

Job applicants should be advised of interview dates, times, and location. It is also useful to review any attire that must be purchased for the position (e.g., a tuxedo), so the applicant is forewarned of any future initial expenses involved in accepting a job offer.

Proofs of eligibility to work (passport, birth certificate, Social Security card, driver's license, etc.) should be listed so applicants will know what documents are necessary to complete the hiring process.

Management Name List

The front desk, banquet department, group sales office, and reservations office should have an up-to-date list of all management personnel, their titles, telephone extensions, and beeper numbers. The daily schedule for the manager-on-duty (MOD) during each time frame should be available in each office area.

Emergency Telephone Numbers

The front desk, banquet/sales office, and reservations office should also have a list of emergency telephone numbers:

- ♦ Police department.
- ♦ Fire department.
- ♦ Ambulance service.
- ♦ Electrician, if not on staff.
- ♦ Plumber, if not on staff.

The Front of the House

The "front of the house" consists of the areas guests are privy to—where they can walk, what they can see. In these areas, the floors must be polished, brass and copper must shine, mirrors must sparkle. Furniture must be dusted daily, carpets cleaned regularly, and wallpaper replaced frequently. Every effort must be made to keep these areas looking extremely attractive. In contrast, the "back of the house," including the kitchens, receiving and loading docks, office space, reservations office, and employee locker areas are often designed with function and sanitation foremost, aesthetic considerations taking second place.

Some of the front-of-the-house areas involving customer contact include:

- ♦ *À la Carte Dining Rooms.* Serving people with reservations as well as walk-ins for dining room service at a table, usually handling groups up to 12 to 20 people (the number is determined by each operation).
- ♦ *Counter Sales.* A subgroup within à la carte sales for guests who sit at a counter rather than at tables and chairs. In a large operation the counter may be physically separate from the dining rooms. Counter service is usually characterized by fast service and frequent turnovers of parties.
- ♦ *Takeout Sales.* Often a subgroup of à la carte sales or counter service offering a limited menu or special menu items for takeout or delivery.
- ♦ *Banquet Department.* Usually a separate department within food and beverage operations that provides total services for a function (e.g., birthday, wedding, anniversary, sales meeting) and involves the total utilization of a physical space (e.g., room). Special menus accommodate customer requests. Often flowers,

special linens, favors, photographers, bands, disc jockeys, etc. are arranged for each function.

♦ *Group Sales.* Sales to a specified number of persons (usually 15 or more), with a preset menu, time frame, and dining room for the meal.

♦ *Gift Shop.* Separate physical space for selling items to restaurant patrons, often including freshly prepared baked goods, jellies and preserves, sausages, pâtes, as well as souvenir items—wristwatches with the establishment logo, umbrellas with the same logo, sunglasses, pens, stuffed animals, and other novelty gifts.

A gift shop provides a great marketing opportunity. Many kinds of souvenirs can be created: notepaper blocks with the logo or picture of the establishment on each side of the block; pieces of the restaurant china, if it has a distinctive pattern; T-shirts or baseball caps with the logo; pens, sunglasses, umbrellas, or scarves with the establishment colors, logo, or motif on them.

A department head or manager is usually in charge of each department. Other key management players could include the following:

1. *Restaurant Directors.* These are the managers of the à la carte dining rooms. In addition, the restaurant directors oversee the service and often manage the counter sales and takeout sales. One or two are usually on duty during a given shift.

2. *Maître d'Hôtel.* Maître d's usually oversee hosts and coatroom personnel as well as service personnel. They are responsible for managing the front desk and the seating and service in a particular à la carte room or rooms. The captain of a service team, often including servers and a busser, oversees the seating and service provided by the team for a specific station within a dining room.

3. *Banquet Maître d'Hôtel.* Most banquet departments have a maître d' who oversees all banquet or private party functions going on at one time. There is generally a captain for each function who oversees all service within that specific function. The banquet department often services meeting rooms as well, setting up for breakfast or luncheon meetings as well as planned coffee breaks.

T w o

Coat Checking

Coat-check personnel are another "first impression" category of employees. If the coatroom is visible from the entrance to the restaurant, guests sometimes will go there before proceeding to the host station.

It is vital that the coat-check attendants be smiling, cheerful, and meticulously groomed because their appearance not only reflects on the operation as a whole but influences how the guests perceive their articles will be treated while in the establishment's care. A sloppy attendant gives guests the impression that their articles will receive sloppy handling.

The purpose of the coatroom is to relieve guests of cumbersome items while they are dining, such as coats, hats, parcels, umbrellas, bags, briefcases, and strollers. The checking of items should be friendly, fast, and professionally efficient. The attendants' return of coats and other checked articles to guests should be equally friendly and efficient, while firmly requiring the submission of proper coat-check stubs. Superior coat-check service can promote the goodwill of the establishment by enabling guests to better enjoy their dining experience.

Since the coat-check personnel may be the first representatives of the restaurant that guests encounter, attendants should be able to answer general questions about the establishment, the menus, the days and hours of operation, and local area attractions.

Often the last employees that guests encounter are also coat-check personnel, so they should have some means of calling a taxi for guests who request this service. Some restaurants have a "taxi needed" call button or switch to illuminate a sign outside the restaurant for passing taxis. Other establishments, not on busy thoroughfares where cabs routinely pass by, may have a direct phone line to a local taxicab company; the touch of a button places a call and a taxi is dispatched for the guests.

Coat-check personnel often report to the front desk maître d' hôtel, who in turn reports to a restaurant director or manager.

The balance of this chapter describes the many facets of a coat-check attendant's duties.

❖ SETUP DUTIES

First check the coatroom for cleanliness and neatness. If necessary, sweep the room or ask the appropriate personnel to do so.

Next check that there are supplies of necessary items:

◆ Coat hangers.
◆ Scotch tape and masking tape.
◆ Pen(s) and paper.
◆ Rubber bands.
◆ Clothespins.
◆ String.
◆ Small pieces of paper, or small tablet.
◆ Small, heavy objects (e.g., paperweights) for anchoring tickets.

Obtain coat-check tickets (with stubs). These are often maintained and monitored by the cashier or manager. The coat-check tickets usually have two parts. One piece has a hole that goes over the hanger and marks the item, or is attached to the item to be identified. The other half (the stub) is given to the guest for redeeming the article(s).

Some tickets have three parts: The top section, with the hole, is placed on the hanger with the coat, the middle section is used for accessory articles (e.g., umbrella, hat, shopping bag, briefcase), and the stub section is given to the guest.

Because every item must have an identification number, paper and pens should be available to make supplemental tickets with the same number on them as on the original ticket. If a guest brings a coat, a hat, a briefcase, an umbrella and a shopping bag, each of the five items should have the ticket stub number on it.

The cashier or manager generally provides an envelope with the tickets, along with a cover sheet identifying the range of numbers, and a coat-check issue control form from the controller or accounting office to control the tickets issued. Check the numbers of the tickets you receive and sign for them. If there are two or more colors of tickets (for different private parties), check the numbers on each color, and sign for those tickets, too. Verify that the stub numbers in your possession are correct; you are responsible for all tickets issued to you. (In a sense, you just "bought" those stub numbers!)

Obtain a bank from the cashier. It will usually be a nominal amount (e.g., $20), sufficient to make change.

If a lockbox is used, pick up an empty lockbox. One lockbox should be used per shift, and it should be placed inside a larger locked cubicle to which only the manager—not the coat-check personnel—has a key. A small slit is needed for depositing money and used ticket stubs, but the opening should be small enough to prevent a hand from reaching through it. The lockbox should also be deep enough that all monies drop securely to the bottom and cannot be easily retrieved.

❖ GENERAL RULES FOR THE CHECKING OF ARTICLES

Coat-check personnel are responsible for all articles placed in the coatroom. While this may sound simple, effective coat-check management needs to be quite organized.

Make sure that all articles are properly checked in with tickets, and do not return any article without receiving its corresponding stub. (The procedure for dealing with lost check stubs is described later in this chapter.)

To avoid having a large amount of available cash at the coatroom, which may invite robberies, all coat-check money and the used tickets and stubs should be dropped into a lockbox, or preferably two lockboxes (one for money, one for tickets), immediately after returning the customer's items.

Tickets cannot be left on the hangers because there would be no "system." When returning a coat or other article to a guest, remove each ticket with the article to avoid confusion when the next article is hung. When the next guest's coat is placed on that hanger, there should be only one ticket—the one for that garment. With the exception of fur coats, articles should be placed in numerical order so they are easy to find.

Coats should never touch the floor of the coatroom, nor should ticket stubs be allowed to drop onto the floor. The floor should be kept clean at all times.

Never leave the coatroom unattended. If you require more monies for your bank, you have several options: (1) Ask someone from the front desk to request that the manager stop by the coatroom, take a large bill (e.g., $20 bill), and procure change from the cashier; (2) take a large bill from your bank and ask a front desk host to go to the cashier for change for you; (3) ask a host to cover the coatroom for you while you procure change from the cashier.

At the end of the shift, when, and only when, the last coat has been returned or, if left unclaimed, moved to the manager's office, complete the cover sheet. On the cover sheet, note how many tickets have been used and how many are being returned to the cashier. Sign the cover sheet and return it, with all unused tickets and the lockboxes, to the cashier or manager.

The cashier will open the lockboxes in your presence, remove the stubs, count the money, compute the number of tickets used, and verify that all is in order. Any tip monies deposited into the lockbox are returned to the coat-check personnel.

❖ TYPICAL COATROOM POLICIES

At a "host-paid" party, where a single customer pays the coat-check charges for all guests attending that private party, a different color ticket should be used for those guests' articles.

The cashier or manager should supply tickets of different colors for each type of function or party. For instance, at a restaurant having three functions—regular à la carte dinner service, a party given by Mr. Smith, and a banquet—yellow, red, and blue tickets could be issued: yellow tickets for regular à la carte dining guests, red tickets for Mr. Smith's party, and blue tickets for the banquet.

When obtaining the tickets from the cashier or manager, you should check the function sheet to determine whether you are charging for all coats that evening, only for à la carte, or for some combination thereof. The function sheet, which lists all parties and banquets for that day and evening, will indicate whether or not they include host-paid checking. (If host-paid valet parking is available, that will also be noted.)

If several coatrooms are being utilized for various parties, the cashier or manager should try to issue different color tickets for each coatroom. If, at the end of the night, the few remaining coats are consolidated in one coatroom, the person on duty, who knows which color corresponds to each party, can quickly service the departing guests. To avoid confusion, irritation, and anxiety among the guests, a prominent sign at the closed coatroom(s) should state where the coats and other articles have been moved.

Host-paid checking services will be noted on the issue control form. Tickets for host-paid parties should be used only for guests at those events.

The "How To" of Coatroom Checking

Hanging the Coats

Carefully take the article(s) from the guest(s). Unless house policy dictates otherwise, each guest is charged the same price for coat-check services regardless of the number of accessory articles. There is usually no extra charge for checking briefcases, umbrellas, shopping bags, hats, strollers, suitcases, and so on, along with a coat. Some establishments do not charge any fee for this service, but permit and even encourage giving gratuities to the coatroom attendant.

When taking coats, if a man and woman are together, reach for the woman's coat first, then the man's. Often the man will hand you both garments. Place the coats gently on the counter or drape the coats over one arm and give the guests their ticket stubs.

If there is a scarf with a coat, place it temporarily in a pocket or an armhole. When inserting it into the armhole, keep the ends visible. When hanging the coat, remove the scarf, place it on the hanger first, then hang the coat atop the scarf, and affix the number to the hanger with the number visible. This method ensures that the scarf won't fall out of the sleeve if the coat is moved.

If several couples arrive at once, your handling of the situation depends on the counter space available and the number of coat-check personnel on duty. If you have little space (the usual situation), take the coats from one couple, give them their tickets, and hang the coats. Then proceed with the next couple or group.

If there is a wide counter and two or more attendants are on duty, take the coats from the first couple, gently lay them on the counter, give the couple their ticket stubs, and place the tickets securely atop the coats so you know which ticket belongs with which coat. Ask the next couple for their coats and repeat. As one person is taking the coats and giving the guests their stubs, the other coat-check person is hanging the garments.

For a party of two or more, double-hanging coats on the same hanger conserves space, enabling you to fit more coats into the same amount of rack space. This technique works only with dry cloth coats, not with fur coats, damp raincoats, down parkas, or other bulky coats.

Every coat gets a ticket so that guests who leave at separate times will each have a ticket. Double-hung coats would have two tickets on the hanger. Otherwise the stub must be marked "two coats," and initialed by the attendant so if another attendant returns one coat he or she doesn't wonder where the "extra" coat came from. Also, it would be inconvenient to the guests. They must both leave together, or at the very least, come to the coatroom together so the number on the stub can be verified, one of the two coats can be retrieved, and then the stub returned to the person who's staying. Don't accept a customer's verbal description of a coat or other article; insist on seeing the ticket stub.

Ticketing Articles

An article means a coat, and any accompanying items, such as the following:

♦ Coat and briefcase.
♦ Coat, hat, and briefcase.
♦ Coat and umbrella.
♦ Coat, hat, umbrella, briefcase, and shopping bag.

If the guest has several accompanying articles, take a piece of paper and improvise additional tickets, as needed, with the same number. These tickets can then be attached so that each article has the number affixed to it. Be sure to do this in such a way that you do not damage the items. Detailed instructions for this procedure appear later in the chapter.

The total number of accompanying articles should be written on the ticket with the coat. This ensures that when the guest returns with the stub, the attendant can quickly ascertain how many other articles were checked with the coat.

Keeping Checked Articles in Order

Place hangers in numeric order to enable fast returns. If two or more coat-check personnel are hanging coats, keep the correct sequence by using the same rack. Fill each rack before moving on to the next one. If you have stationary poles for hanging coats, begin hanging coats at the back and work toward the front. It may require extra work at first, but it really is the most efficient method. As coats are received and hung on hangers, you can quickly hang the most recent in front, pushing the stack of previously hung coats to the back.

When there is a "break in the action," move the hung coats as far back as possible, leaving room in the front of the room for the next group of arriving or departing guests.

If there is a "moving chain" coatrack, placement is not a problem. As the spaces in front are filled, the chain moves the hung coats toward the back, opening up new spaces in the front.

The ticket number should be visible at a glance; most tickets are printed on one side only, so make sure the printed side is visible to facilitate quick returns. If double-hanging coats, use one ticket per coat. If two coats are on one hanger, then there should be two tickets on that hanger.

Dealing with Awkward Articles

When attaching tickets to accompanying articles, follow these guidelines:

♦ *Umbrellas.* Wooden clothespins are recommended for affixing tickets to umbrellas. Attach the stub or slip of paper with the number to one of the fabric or plastic panels. Store umbrellas, collapsed, in an umbrella stand or large clean plastic tub (you can usually get these from the kitchen).

♦ *Hats.* The ticket's corner should fit behind the ribbon circling a hat, fitting between the ribbon and the hat. Be sure the number is visible. Place the hat atop a shelf (directly above the coat, if possible) so nothing is placed on top of it and it will not be crushed. If the hat has no ribbon, flowers, or feathers suitable for holding the ticket, place the stub on the brim, resting up against the hat so you can see the number just by glancing at the shelf. Or, tape the number to the shelf, with the number hanging down so it's easy to spot, and place the hat above it.

♦ *Strollers.* Use a clothespin to affix the number to the stroller.

♦ *Briefcases.* If briefcases are being placed on the floor, stand the briefcase on its bottom, handle up (so it takes up minimal space). If possible, slide one corner of the ticket between the two sides (left and right) of the case. If it won't stay in place, lay the stub atop the briefcase and anchor it there with a small, relatively heavy item such as a paperweight.

If storing briefcases on a shelf above the coats, stand the briefcase upright, directly above the coat, or as close to the coat as possible. Tape the number to the shelf and place the briefcase directly above the number with the narrow end facing out.

Be careful when using tape on items, especially on leather or gold-plated metallic trim, which is often found on briefcases.

You may pull off the finished top layer of leather or the plating metal when you remove the tape.

♦ *Shopping Bags.* Follow the procedure for briefcases, making sure the narrowest side is facing out. This will give you maximum use of shelf space. If bags are stored on the floor, use a clothespin to affix the number to either the shopping bag handles or one side of the bag.

Fur-Coat Policy

When a guest approaches with a fur coat, take the coat as usual and ask for the guest's name. If they don't want to divulge their name, for whatever reason, ask the person to supply a fictitious name. Explain to the guest that he/she will be asked for by this name when he/she returns for the coat, as a safety precaution. Explain this procedure in a low voice so you are not overheard. A person standing nearby should neither be able to hear the name given nor see or ascertain the coat-check number given to the guest. Write the name the guest gives you on the back of the coatroom ticket, and keep it out of sight when hanging the coat.

Some operations will have separate "fur coat" tickets, labeled "F1," "F2," and so forth. Using separate tickets will prevent people standing in line from determining that the fur coat stub was one less (or two less, etc.) than their numbers, since most tickets are issued in sequential order for accounting purposes. Fur-coat tickets are issued by the cashier or manager and signed for by the coat-check personnel in the same manner as other tickets.

If you have heavy-duty hangers, use them for the fur coats, as these coats are heavy and bulky. Hang the furs in the coolest part of the coatroom—furs do not like heat. Don't squash fur coats together; if possible, leave some space between them so they can "breathe."

When the guest returns to claim the coat, ask for the name that was given when the coat was checked (the name that should be on the back of the ticket). Take the guest's numeric stub, find the garment, verify that it is in fact a fur coat, (sometimes stubs get exchanged or mingled by guests), and check the name written on the back of the ticket.

If the name given by the guest matches the name on the back of the ticket, remove the coat and the ticket from the hanger, verify again that the name given to you when the coat was checked is the same as the one the guest has told you, then return to the front, and give the guest the coat.

If the name the guest has given does not match the name on the back of the stub that is hanging on the coat, call a manager. Do not return the coat to the guest. The stub may have been lost (e.g., the stub may have been left in the restroom, or a purse may have been stolen), and a person who is not the rightful owner may be attempting to reclaim the coat. The manager must now take steps to ascertain rightful ownership. (The general procedures for dealing with lost check stubs are detailed later in this chapter.)

While these may seem, at first, to be lengthy steps to go to for a fur coat, guests actually prefer that these security measures be taken and are usually more than happy to cooperate. Most guests want to do everything possible to ensure the safety and well-being of their property. By taking the aforementioned steps, coat-check attendants are definitely going beyond the basics!

Expected Degree of Care

The establishment and its staff is obligated to treat any and all checked articles with, at a minimum, ordinary care—the amount of care that a person would ordinarily use considering the circumstances, to preserve the articles.

This definition of care means that you cannot allow coats to brush along the floor, or have scarves drop out of pockets or sleeves onto the floor, or drop briefcases or shopping bags. Carry heavy shopping bags from the bottom, not using the handles, so the bags won't rip or tear while in your care.

When attaching a ticket to an article, you may use any means that will not damage the article—tape, rubber bands, clothespins— but exercise extreme care not to mark or puncture the item.

Returning Articles

When a guest presents a check stub(s), take the stub(s), find the article(s) that those stub(s) represent, and return all articles and accompanying articles to the guest.

Lost Check Stubs

Sooner or later, you will encounter the guest who has lost his or her check stub. A standard procedure should be developed and used whenever this situation occurs.

Following the prescribed procedure ensures that if, later, a guest arrives with the missing ticket stub, you can obtain the same information from the guest with the stub and identify the correct owner. The stub might have been lost and someone has now picked

STEPS IN RETURNING ARTICLES

1. Take the stub and collect the coat-check fee from the guest, unless the house policy is to have the guest pay when checking the articles, or the charge has been paid by the host of a private party or banquet.
2. Locate the hanger. Pull the ticket off the hanger, and remove the coat and scarf, if any. Check the ticket for a handwritten number to determine if there are any additional items.
3. Find and gather all additional items. Remove the ticket numbers from all articles of the guest and return them along with the coat.
4. Deposit all sections of the ticket and monies in the lockboxes.

Reminder: Be sure to pull all tickets from the hangers when removing coats to avoid confusion when reusing the hangers. When the coat is gone, so should be the stub. The hanger should be empty, the floor clean (never drop stubs on the floor), and all stubs and money safe in the lockboxes.

it up and is trying to claim the item(s); or the first person might have mistakenly claimed item(s) similar to his or her own.

Be alert to any unclaimed items at the end of the evening. There might be a coat or other item (e.g., briefcase), identical to the one given to a guest who lost a claim ticket. If two or more coats or other items are nearly identical, you'll need the lost guest-check form to trace the person to whom the other article(s) were returned.

Never allow customers behind the counter. They'll say, "But I can find my coat in a minute . . ." or, "But I'm a senator and I'm late . . ." It doesn't matter who they are, or how late they are, do not allow customers behind the counter.

Closing the Coatroom

It is first necessary to calculate the number of tickets used:

♦ Determine the next ticket number that will be used.
♦ On the cover sheet, find the stub number for the first ticket used.

LOST CHECK STUB PROCEDURE

1. Ask the guest to identify the coat or items checked. The item(s) must be identified not only by visible identifying marks that someone could have observed while standing near the coatroom (e.g., red wool coat) but by some hidden identifier—a loose button; the pattern of the coat lining; some item in a pocket, the briefcase, or the shopping bag; the hat size—something that would not be apparent to someone who might have been standing next to or behind the person checking the item.
2. Find the item(s). Do not allow guests into the coatroom to locate a "lost" article(s); they might "identify" items that are not theirs, but are similar, or they might even locate items that are not theirs but that they'd like to own.
3. Check the identifying description against the actual article(s).
4. If they match, fill out a lost guest-check form, which should include the following information:

    Name, address, and phone number of guest, verified by driver's license or other identification form; include both home and hotel telephone numbers of an out-of-town guest.

    Description of item(s), including the hidden identifiers.

    Three pieces of identification (driver's license, Master-Card, Visa, Amex, etc.). At least one should have a photo.

    The guest check ticket stub number—obtain this from the actual article(s), once you've located it/them.
5. Return the identified article(s) to the guest.
6. If the description and the item you've located do not match, call the manager. Do not release any items to the guest.

♦ Subtract:

    Next number to be used
    − First number used
    ────────────────────
    Total number of tickets used

♦ Enter this number on the cover sheet.

Ask the manager on duty to unlock the outer lockbox. Take the transportable lockboxes, cover sheet, control form, unused tickets, and bank to the cashier or manager.

The cashier or manager will open the lockbox, reimburse the house for tickets used, reimburse the house for bank monies issued, and return all other monies (gratuities) for distribution among the coat-check personnel who worked that shift.

All policies and procedures should be followed to the letter. Management should inspect the coatroom regularly, to make sure that every article has a ticket and that only tickets for a given shift are being used. Breaches of regulations usually warrant disciplinary action.

T h r e e

Lost and Found

It happens. Guests lose items—sunglasses, cameras, shopping bags, purses, jewelry, wallets, combs, brushes, notebooks. Sometimes there are identifying items, such as a monogram or a phone number, but many times there are not.

Each operation should establish lost-and-found procedures and disseminate those procedures among the staff so everyone is aware of what is to be done with found articles as soon as they are discovered.

Customers who have lost an article and then have been able to reclaim it, whether it's later that same day or a few days later, will remember that establishment with appreciation. They will recall the concern expressed by the staff while filling out the Lost Item Report and while searching for the lost article.

❖ ESTABLISHING STANDARD PROCEDURES

Lost-and-found procedures are among the difference makers that can determine whether or not guests will return to your establishment: Did the staff seem friendly and concerned about the lost item? Did they help search for the lost item? Did they take sufficient information so that the item, if found at a later date, could be returned to the owner?

Basic procedures follow. They can be modified as desired, but if guests can answer the preceding questions in the affirmative, then the procedures have accomplished what they set out to do.

Guests usually notice the loss of an item after leaving the establishment and typically will call or stop by to ask if the lost item has been found.

In either case, an employee must obtain and record a description of the lost item, and verify identification of the owner. Part

One of the Lost Item Report should include the following information about the lost item:

♦ A physical description, including size, shape, and overall condition.
♦ Any identifying marks, such as engraving, embroidery, appliques, paint chips, scratches, or stains.
♦ Approximate age of item.
♦ When noticed lost.
♦ Last known physical location of item (e.g., at the table by the window in station 2 (table 42), in the ladies' room, or on the lobby ledge).

Part Two of the report, the owner identification, should include the following:

♦ Name, home address, home and business phone number (or whatever contact number is supplied by the person).
♦ Hotel address, phone, and date of departure if a visitor to the city.
♦ Three forms of identification (e.g., driver's license, two credit cards).

All items found in the à la carte dining rooms during dining hours should be given immediately to the front desk maître d'hôtel or other manager on duty. Any items found in the banquet or private dining areas should be given to a manager in the banquet department.

All cash and other recovered items of significant value (wallets, purses, and jewelry) should be kept in a locked location for instance, in the managers' office. Small items of nominal value (hats, scarves, sunglasses, costume jewelry, etc.) can be held in a locked drawer at the front desk. Regardless of where the articles are found, they always should be stored in locked locations that management can easily access when there is an inquiry. If the only manager who has a key to the lost-and-found happens not to be on duty when a guest inquires about an item, the resulting inconvenience does little to enhance the guest's image of the establishment. An employee with access to the lost-and-found must be available during all hours when the establishment is open to the public.

Without calling a manager, the maître d' at the front desk can return small items to their owners after obtaining proper identification. Larger items, such as hats and coats, should be kept in a locked closet or other secured area, such as a designated locker. If items are very large, they might be kept in the manager's office.

Generally, all lost-and-found inquiries should be directed to the front desk maître d' or the manager on duty. If there is a separate banquet department, however, banquet/meeting room losses should be handled by that department. In such establishments, the first question to inquirers should be, "Did you lose the item while dining, or at a private party or meeting?"

This enables one to direct the call to the proper area and will save the guest time in ascertaining whether the lost item has been found.

As soon as an item is found, a manager should check for any identification. In the case of a purse or wallet, the manager should, in the presence of another management-level employee (at least two people should be present), look through the item for identification and the owner's phone number so the person can be notified and can arrange to retrieve the article. If there is cash in the wallet or purse, the amount should be noted and initialed by the managers as part of their nightly report, as well as whether they were able to contact the owner and whether/when the person will be returning to pick up the item. At the owner's request, the item may be returned via mail or messenger service. When mailing the item to the owner, the establishment should insure the item and send it return receipt requested.

Every operation establishes policies for the disposition of unclaimed items. Often, umbrellas and small articles, such as sunglasses, are left at restaurants and never reclaimed—perhaps guests just don't remember where they last used them. Many restaurants have a "30-day holding period," after which the articles are given to a local charity.

❖ RETURNING LOST ARTICLES

A person who inquires about a lost article, should be asked whether he or she has completed a Lost Item Report. If so, briefly ask the person to describe the item, when the loss was reported, and to whom. Locate the report and match the current description against the written description.

Usually it's quite simple to find the report; the person usually states that he/she reported it to Mr./Ms. Chelsea, who was the manager on duty last Saturday night, and so on. If it's important enough to the person to stop by the establishment or call, the item is important to them and they'll remember when they reported it, to whom, and so on. If no report has been made, complete one.

With the report in hand, check the lost-and-found depositories to determine whether the item has been found. If it has, verify owner identification. If the identification of the person who reported the missing item and the person who is presently claiming the article match, return the article, mark the report, "Item Returned" (or check the appropriate box on the form), and note the date the item was returned. Both the person claiming the item and the manager or maître d' returning the item should sign the form, noting that the guest/claimant has accepted delivery of the item and that the manager has released the item to the claimant. These signed reports should be saved for at least 90 days (preferably for one year) in a "Returned Items" file, to avoid any disputes about "mistaken returns."

If the owner verification on the report does not match the identification of the person who is presently in the establishment, a manager should be summoned. It is possible that two people lost, or forgot to leave with, their size 10 Burberry plaid trench coats at the end of the same evening. Perhaps the two coat owners were at different parties or perhaps they were at the same large fund raiser. If that is the case, management must make every effort to return each coat to the rightful owner.

Consider this scenario, which has, in fact, occurred. Two identical coats are checked one evening. One has empty pockets; the other has a scarf in a pocket. At the end of the evening, a guest is about to leave, but cannot find her coat-check stub. With a description of the coat, the lining, and the type of material, the attendant determined that the coat with ticket 436 hanging on it was the coat in question and returned it to the guest. The owner vaguely thought she had placed her scarf in the coat pocket, but after all, that was hours ago, and she'd had several glasses of champagne. She must have left the scarf at home.

About an hour later, another guest plans to leave, and presents coat check stub 436 to reclaim her coat. At this point, it might seem it should be obvious to the coat-check person what has happened. However, several coat-check persons were on duty, the coats from two or more coat-check areas have now been consolidated into one coat-check area, and the attendant who dealt with the first woman went home early.

In any case, coat 436 could not be found. It was finally determined that it had already been returned. Also, a virtually identical garment was located, and management discovered the Lost Check Stub report showing that "coat 436" had been given to another guest. The "436" coat had to be retrieved for its rightful owner, and the remaining "lost" coat, returned to the first woman, an awkward situation for the management. When such mix-ups occur, the establishment usually will send a courier to deliver the coat that was left in the coatroom. At the same time, the courier will pick up, then deliver, the other coat to its owner.

A paper trail is essential because guests may take days to notice something is missing, especially when they simply walk out of parties and forget to get their coats. It might have been a mild, rainy evening when guests arrived, several wearing raincoats. At the conclusion of the party, the weather was clear, and several guests, in the midst of conversation, walked out together, perhaps called a cab or got into one car, and proceeded to their next destination. The next time it rains or gets cool, they look for their coats and can't find them, and finally try to remember where it was that they last wore a raincoat. They may or may not call.

Anything you can do to assist the guests with any of their problems will be remembered. And if it's a favorable memory, when it's time to choose a restaurant or theme park or entertainment center, they'll remember your establishment and return! And that's what makes the difference.

_F o u r_

Guests and Employees with Disabilities

It has always been important for business owners to take care of the needs of guests and employees, and to ensure their safety and safe passage from one part of the building or grounds to another. It is equally important to include those with disabilities in all activities, including employment. This philosophy reflects not only good business sense, but a moral and civic duty for establishments that want to be good community citizens, going beyond the basics.

❖ THE AMERICANS WITH DISABILITIES ACT

Civil rights legislation, in conjunction with the passage of the Americans with Disabilities Act (ADA) has far-reaching ramifications for operators of restaurants and other establishments open to the public. With the ADA taking effect for most properties on July 27, 1992 (employers with 25 or more employees), and for others in 1994 (those with 15 or more employees), owners are now legally bound to ensure equal access to their facilities, activities, and employment, and to provide reasonable accommodations for those with disabilities. To be profitable, in today's world, businesses must utilize every available marketing advantage and difference maker that they can, including making the establishment "disability friendly."

There are many forms of disabilities. Some people are vertically challenged, perhaps by their use of a wheelchair, and so need to have phones and elevator buttons at a lower height. Some people are hearing impaired; others are blind or visually impaired.

It is relatively easy to provide auxiliary aids and services to guests and employees with vision or hearing impairments. For instance, if a convention is meeting at the property, arrangements should be made for a person to sign the speech so hearing-impaired visitors can follow the proceedings. Any and all printed handouts should be available in braille. Phones in rooms can be equipped with telecommunications devices for the deaf (TDDs).

Adjustable Lecterns

Modern lecterns can adjust to accommodate people of varying heights, in either a sitting or standing position. Think of a box with a deep-sided lid. Raising the lid to its highest position creates a tall box; sliding the lid down over the lower half makes a shorter box. This kind of lectern is designed to rise quietly from approximately 37 inches (a comfortable height for someone in a wheelchair), to 47 inches full standing height, or anywhere in between. This lectern will be equally comfortable for a child speaker, a vertically challenged person, or a tall person. No more standing on boxes! And the speaker opening is wide enough to accommodate a wheelchair (40 inches). Figure 4.1 shows the VanSan Corporation's Omni model adjustable lectern, which is currently available.

FIGURE 4.1   VanSan "Omni" lectern.
(Courtesy of VanSan Corp.)

Language Services

Today, more so than ever, we are living in an international market-place. It is not unusual to be calling for parts or equipment to an international vendor; it is common for guests to be arriving from various continents, many countries. With airplane travel wide-spread, people arrive from all over the world at virtually all tourist destination spots, whether for relaxation, a convention or confer-ence, or an expedition. In addition, "signing" has become recog-nized as a true language, of vital importance to many individuals.

It is important to truly welcome guests and make them feel very special when they arrive. One way to do that is to ensure that someone on the property speaks their language. Many establish-ments place a premium on employees who can speak two or more languages, including signing. One property at a popular tourist des-tination gives employees fluent in a second language a 25-cents-per-hour differential in their pay rate with 10-cents more per hour for every additional language spoken fluently.

These employees typically hold guest-contact positions such as receptionist at the front desk, tour guide, reservations clerk, server in a restaurant or concession stand, or bellhop or valet parker. Someone who can sign is also in demand for translating speeches at meetings and conferences. He or she often stands on the stage or at the front of the meeting room along with the guest of honor and other speakers.

Ingress and Egress Accessibility

Elevators are being updated with reasonable accommodations for the disabled. Elevator buttons are located at a height enabling peo-ple in wheelchairs to operate the elevator. Floor numbers may be printed on the button and also in braille and/or raised numerals next to the button. As the elevator moves from one floor to an-other, a chime may sound, so the user can count the floors. Some electronic "talking elevators" announce the floor number as floors are passed or announce the floor number when the doors open.

Physical barriers must be removed or alternative methods pro-vided to reasonably accomplish the goal of "accessibility for every-one." As an example, ramps should be provided as well as stairs so those using wheelchairs, or crutches, or who are otherwise unable to or prefer not to utilize stairs will have access to all areas. This is especially crucial in the parking lots, adjacent sidewalks, and en-trance areas.

DISABILITY-FRIENDLY ❖ ESTABLISHMENTS

Disability-friendly establishments try to enable people with disabilities to feel comfortable while on the premises. Different establishments have different circumstances. Some are historic and must balance the preservation of artifacts against modernization to accommodate the disabled. Some establishments are carefully budgeted and cannot undertake extensive, and expensive, renovations. The following sections explain modifications that will enable people with disabilities to overcome those problems to some extent. Some are optional; some legislated, such as signage requirements for meeting specific government-established standards. Some ideas can be adopted or incorporated without much expense or renovation, such as installing TDD equipment; others require a substantial capital investment, such as replacing stairs with ramps or adding ramps and/or elevators.

Public Telephones

Ease of Access

Telephone stations have been redesigned to better serve people's needs. More and more phones in hotels and restaurants are countertop models, with a bank of several phones along a wall, above a writing counter. They are approximately 4 feet from the floor and have no booth dividers, allowing guests in wheelchairs easy access. Some upscale hotels and resort hotels provide private telephone booths that can accommodate a wheelchair.

Devices to Assist the Hearing Impaired

In addition to recent changes for the vertically challenged, there are advances for those who are hearing impaired, namely telecommunications devices for the deaf (TDD). These usually are available in two models: display-only models that display the words as the person on the other end of the line types them in, or display/printer models, which both display the message and type it out on a printer at the receiving end. These can be mounted in public spaces or attached to hotel guest room phones.

Restaurant guests may request TDDs, and several models are available to accomodate them. In addition to those permanently mounted on a shelf or wall for connection to a public phone, others are portable and can be used for guests in various locations.

A TDD is basically a three- or four-row keyboard, just like a typewriter or PC keyboard, with a special telephone handset receiver at the top of the keyboard. When a call is made, the conventional handset is set into the TDD handset receiver and the message is displayed rather than heard. Standard models have a display panel for the incoming message ranging from 20-character displays to two or even four rows of 40-character displays. Some TDD models have computer chips with 2K to 32K of memory and offer many options, such as the following:

♦ Automatic answering, recording, and storage of messages.
♦ Keyboard or memory dialing.
♦ Directories of frequently called numbers for memory dialing.
♦ Remote message retrieval.
♦ A signal indicating the dial tone, a busy signal, and the ringing of the other phone.
♦ A personalized greeting and message alerting the caller to use TDD.
♦ Time and date of message recorded.
♦ Built-in printer or a connection/port for attaching an external printer (often a portable model feature). Some printers have various size fonts for small to large printing; some have both upper- and lower-case letters.
♦ Phone strobe light that flashes when the phone rings.
♦ Rechargeable battery (portable models).

Public access TDDs are often "ruggedized," that is, enclosed in hardened plastic or stainless steel housings with sealed, pressure-sensitive keyboards that won't get clogged when drinks or foods are spilled on them. Some models can be mounted directly onto an existing telephone shelf; other ruggedized models, embossed with a TDD pictogram, are mounted to the side of an existing phone installation. If the TDD is enclosed in a ruggedized case, the user must first pull down the TDD, switch it on, place the handset of the public phone into the acoustic cups and then make the phone call. If the TDD is mounted onto an existing telephone shelf, the caller simply places the public telephone handset into the acoustic cups and proceeds as usual to make the TDD phone call. Some TDDs are programmed to announce to the called party, "This is a TDD call, so please use a TDD," or words to that effect.

For hearing-impaired guests who require amplification of the voice coming over the telephone, there are adjustable volume control amplifiers that attach directly to the handset cord or onto the phone handset. They can amplify the incoming voice by as much as 20 decibels. Some establishments switch out the standard handset and replace with an amplification handset, which resembles the standard model but has an adjustable dial in the middle that can be rotated to amplify the incoming voice.

Hotels and motels often use visual alerting devices with receivers or signalers for hearing-impaired guests. The receiver or signaler is plugged into a wall outlet and a lamp, usually a bedside lamp, is plugged into the receiver. When there is a knock on the door, the lamp will flash, blinking on and off. When the phone rings, the lamp again flashes, but with a different signal to distinguish it from the door knock. When the clock alarm goes off, a third pattern of flashing light signals is emitted from the lamp. If the smoke detector goes off, there will be another pattern of flashes. Some hotels use a strobe light for better visibility from any area of the room. Some hotels have not only a lamp or strobe light affixed to the receiver/signaler but also a bed vibrator for awakening a sleeping person to a phone call, clock alarm, or activated smoke detector.

Many establishments are installing remote audible and visual fire signals. These devices both sound a loud audible alarm and flash a piercing strobe light when the smoke detection devices are triggered.

To accommodate hearing-impaired guests who wish to watch television programs or videotapes, closed caption decoders are available to access shows offering the captions. Often, a remote control is included so the viewer can scan for a preview of active channels and locate the special text channel that lists the current captioned programs. Viewers using the decoder can watch any channel, although not all shows feature closed captions.

Guidelines for Signage

All signage should incorporate strong contrast of light and dark colors for letters and background. A matte or semigloss finish enhances visibility. Often brushed brass or aluminum is used, with black lettering.

Major Area Signs

Permanent room and space identification signs have specific requirements. The words should be plain, clear, and all in capital

letters. The *sans serif* style of lettering, because of its lack of orna-
mentation, is functional rather than decorative and is very easy to
read. Raised letters should be a minimum of $1/32$ inch in thickness.
The height of the letters should range from $5/8$ inch for smaller
signs to 2 inches for signs used in meeting rooms. Distinctive letters at least 3 inches high should be used for major area signs,
which are usually placed above eye level. No braille or raised letters are required for such signs. The size of the letter should be
dictated by function: How far away is the typical guest who attempts to read the sign? Management must make sure the sign is
readily visible and legible at that distance.

Lettering Style

Directional and informational signs, such as those giving room
checkout times or exit routes in case of fire, may be both upper-
and lower-case letters. The style of lettering determines the width
of the characters, but they should have a clear, readable, and simple
design. Neither disproportionate tall lettering nor wide lettering is
acceptable. The width-to-height ratio for each letter should be in
the 1:1 or 3:5 range. The most easily readable lettering is often in
the 3:5 range, or slightly taller than wide.

The width of each stroke of the letter is important for directional and informational signs. Each segment must be of sufficient
thickness to produce a sharp foreground-to-background image that
is easy to read. There should not be too much background (strokes
are too thin) or too much foreground (strokes are too thick). Acceptable ratios for lettering range from 1:5 to 1:10. With 1:5 lettering, the width of each stroke of the letter is $1/5$ of the total height of
the letter; with 1:10 lettering, each stroke of the letter is $1/10$ of the
total height. Another way to think of it is the width of each stroke
is $1/5$ or $1/10$ of the total height of the letter. For example, if a letter
"H" is 1:5, the vertical lines of the "H" might be 5 inches high by
1 inch across, with the crossbar being 1 inch high.

A sign on a wall that identifies a room or space and is within
easy reach of a guest (e.g., room numbers, elevator buttons), must
also be in braille or raised letters. Braille or raised lettering is not
required for overhead signs, nor for temporary signs. For instance, a
meeting room name or room number would be considered permanent and should be identified with braille or raised lettering. However, the name of the group occupying the meeting room for that
day or that week can be printed in standard format. When braille is
used, it should be grade 2, which includes 189 contractions and
short-form words.

Pictograms

Pictograms, another kind of signage, are symbols within a square background field. Most of the symbols are patterned after the Department of Transportation signs, which have acquired nearly universally understood meanings. Examples include an outline of a woman wearing a dress to designate the women's restroom, an outline of a man wearing a suit to designate the men's restroom, a telephone handset to designate a public telephone, a cigarette inside a circle with a line drawn through it to designate a no-smoking area, and so on. Each pictogram should be in contrasting colors and a minimum of 6 inches high. A verbal and braille translation of the pictogram must be on the sign, typically directly below the pictogram (see Figure 4.2).

Clearance Requirements

Room signs should be placed on the wall adjacent to the latch side of the door or opening. The centerline of the sign should be 5 feet (60 inches) from the floor. In narrow hallways, especially those with only the door at the end of the hallway, or double-doors at the end of the hallway, there often is no adjacent wall to post the sign. In this case, the sign should be placed on the nearest wall. There must be a minimum of 3 inches of clear space from the door swing (path the door takes from being closed to fully opened) to any object or sign, excepting doorstops, handrails, and so on. This ensures

FIGURE 4.2   "No Smoking" pictogram.

that a sign will not be covered if someone opens the door and holds it open.

For an overhead sign, the clearance should be a minimum of 6 feet 8 inches (80 inches). Since this is a standard door height, most overhead signs already meet that requirement.

Appendix A provides examples of signage graciously provided by CAS The Sign Systems Company. For further information, review the appropriate *Federal Register* or contact CAS The Sign Systems Company, Sun Valley, California (818-768-7814), or other signage companies.

Dining Areas

Dining room managers must be aware of space requirements for accommodating guests in wheelchairs. Typically, the bottom of the table needs to be at least 27 inches from the floor (ideally 29 inches), and the table should have an unobstructed width under the table of 30 inches. There should be a minimum of 19 inches of leg clearance under the table, beginning with the edge of the table and going toward the center. The seated person using a wheelchair requires 30 inches from the edge of the table extending outward into the aisle. (A dining room setup usually allows 24 inches for a seated person; an allowance must be added for the wheels of the chair.) Freestanding tables in the dining area usually have a minimum of 6 feet between them to allow an aisle for people to pass through when patrons are seated at those tables. Passage of a wheelchair requires a 7-foot aisle and is highly recommended for at least one aisle of the dining room.

The ADA requires that a minimum of 5 percent of the fixed seating capacity of a dining room be able to accommodate people with disabilities if modifications to do so are reasonably achievable. For new construction and extensive renovations, meeting these goals should be possible and even surpassable.

❖ POSSIBLE TAX CREDITS

Tax credits may be available for establishments that comply with the Americans with Disabilities Act of 1990 if reasonable and necessary expenses for the facilities were incurred to bring the establishment into compliance.

Costs to remove barriers that prevent the establishment from being accessible and/or usable are included as well as costs of delivering materials to individuals with hearing and/or visual impairments and costs for equipment and/or devices such as TDDs (or modifications to such) for individuals with disabilities. Other examples include renovations to the parking lot to create spaces close to the building that are sufficiently wide and level for people with disabilities, and ramps to gain access to the establishment.

The credit is a percentage of the expenses, up to a cap amount for any given year.

❖ # GUIDELINES FOR ACCESSIBILITY

The Accessibility Checklist at the end of this chapter can be used to determine an establishment's accessibility. This list was derived from a study designed by Ricardo Oliver, Accessibility Director, The Center for Independent Living, Orlando, Florida.

The checklist should be used in conjunction with (1) the ADA Accessibility Guidelines (ADAAG); (2) the accessibility requirements manual for your state, if one has been issued (in Florida, a person would access the Florida Accessibility Requirements Manual, or FARM); and (3) the federal regulations regarding Title III of the ADA. The following checklist is not intended to be a substitute for the ADAAG; FARM, or other state requirements; or federal regulations. Its intended purpose is to help users gain a basic understanding of the facility compliance issues of the ADA. These materials are provided for informational purposes only, however, and are not to be relied on in determining such compliance. Legal counsel or establishment management, or both, should be consulted to determine how this information relates to each establishment and its unique situations.

ACCESSIBILITY CHECKLIST

Parking

____ Parking is available near the entrance (within 200 feet)?
____ Parking is reserved for handicapped?
  ____ International Symbol of Accessibility is at each space?
  ____ How many spaces?
  ____ How many covered HP (handicapped parking) spaces?
  ____ How many spaces with 114 inches or more (wide) for van clearance?
  ____ How many covered HP spaces with 114-inch clearance?
  ____ How many spaces 12 feet or wider (8 feet for vehicle; 4 feet aisle)?
  ____ How many covered HP spaces 14 feet (168 inches) or wider?
  ____ How many HP spaces parallel to the curb?
  ____ Curb ramp is easily accessible?
  ____ How many spaces to pass before access to curb ramp?
  ____ Parking is on level ground (slope is less than 1:20)?
____ Valet parking is available?
  ____ What hours?
  ____ Drop-off point is near the entrance (within 100 feet)?

Sidewalks

____ Sidewalks are at least 36 inches wide?
____ Sidewalks are in good repair?
____ Sidewalks are continuous, without interruption?
____ Walkways have nonslip surface?
____ Walkways blend into common level if they cross a driveway?
____ Gradient is not greater than 1:20?
____ At least one accessible route is throughout all parts of the facility?

Curb Ramps

____ Curb ramps are available?
____ Curb ramp is provided where accessible route crosses a curb?
____ Curb ramp is 36 inches wide or greater?
____ Curb ramp has flared sides with a maximum slope of 1:10?

Ramps

____ Signage is available indicating location of ramp access?
____ Signage is high enough to be visible when vehicles are parked in front of or near sign?
____ Access to ramps is unblocked, without crossing curbs?
____ Slope is 1:12 (5 degrees) or less?
____ Handrails are 30 to 34 inches high?
____ Handrails extend 1 foot beyond top and bottom of ramp?
____ Ramp surfaces are nonslip?

_____ Clearance at top and bottom of ramp is 5 square feet?

_____ Ramp has walls, railing, projecting surface or curb at least 2 inches high to prevent slipping off ramp?

Entrances

_____ At least one public use entrance is accessible to wheelchairs from an accessible route (no curbs, stairs)?

_____ There is an entrance from the garage for wheelchairs?

_____ Public transportation is available?

    _____ Bus?

    _____ Cabs?

    _____ Subway/trolley/monorail/train?

_____ What is the distance to bus stop or other transportation?

_____ Doors are electronic or power-assisted?

_____ There are standard single or double-leaf hinged doors?

_____ Force needed to open exterior door is 8 lbf or less?

_____ Force needed to open interior door is 5 lbf or less?

_____ Doors can be opened by a single effort?

_____ Doors are push/pull or lever operated?

_____ What is height of door hardware?

_____ Speed of door closure is 3 seconds or greater?

_____ Doorways have a clear opening of 32 inches wide or greater?

_____ There is a vestibule at entrance?

_____ Doorsill has no sharp inclines or abrupt changes?

_____ Is there a 5-foot-square platform if entrance is ramped and the door swings out?

_____ Is there a 3-foot × 5-foot platform if door swings in?

Lobby Area

_____ Is a portion or all of the registration desk at a 36-inch height?

_____ Are handicapped restrooms available in the lobby area?

_____ Identification signs are in braille or raised lettering?

Elevators

_____ Doors close slowly enough to facilitate entering or backing out if a person is a slow walker, in a wheelchair, on crutches, etc.?

_____ Space is sufficient for those with a walker, wheelchair, on crutches, etc.?

_____ Controls are accessible to those in a wheelchair?

    _____ What is height of controls?

_____ Controls are in braille or raised lettering?

_____ Chimes or verbal indications of floors are used?

_____ Telephone is accessible for emergency use?

    _____ What is height of telephone?

Public Telephones

____ At least one phone is on an accessible route?
____ Clear floor space in front of phone is at least 30 inches × 48 inches?
____ Height of dial is no higher than 48 inches from the floor?
____ Coin slot is no higher than 48 inches from the floor?
____ Telephones are equipped for persons with hearing disabilities?
____ Push button dialing is available?
____ Visual display is available?

Signage

____ Check for directional signage to:
    ____ Telephones?
    ____ Restrooms?
    ____ Meeting rooms?
    ____ Restaurants?
    ____ Lounges?
    ____ Snack bars?
    ____ Game room?
    ____ Pool?
    ____ Gift shop?
    ____ Office?
    ____ Front desk?
    ____ Lobby?
    ____ Bedrooms?
____ Identification signs are in braille or raised letters?
____ International Symbol of Accessibility or directional arrows identify accessible public restrooms?

Water Fountains

____ Are water fountains on an accessible route?
____ Persons in wheelchairs can wheel up to fountain?
____ Fountains use single hand operation?
____ Operating controls are designated with braille or raised lettering?
____ Up-front spout and controls are 36 inches from the floor?
____ Clearance under fountain is 27 inches?

Warning Signals, Exit Signs

____ Are they:
    ____ Visual?
    ____ Audible?
    ____ Clearly audible above prevailing sound level?
    ____ Flashing and/or illuminated?
    ____ Exit signs are in green?

Meeting Rooms

____ How many meeting rooms are accessible by disabled?
____ There is an amplified sound system?
____ Ramp leads to elevated stage or speaker's platform?
____ Restrooms are accessible from meeting rooms?
____ There is a vertically adjustable lectern?

Accessible Routes

____ Are there accessible routes to:
    ____ Pool areas?
    ____ Exercise rooms?
    ____ Game rooms?
    ____ Gift shops?
    ____ Restrooms in areas other than lobby?
    ____ Office?
____ Is lift-equipped transportation available throughout facility?
____ Are rental cars available?
    ____ Are hand controls available?

Dining Areas

____ Seating is accessible, including accessible route to seating?
____ Floors have nonslip surface? What type?
____ Table height is 29 inches, floor to bottom side of table?
____ Aisle space between tables is minimum of 30 inches, preferably 36 inches?
____ Aisle tables have 5-foot turning radius (or T-shaped clear area at dead-end tables)?
____ Movable seating is available?
____ Portable raised leafs are available?
____ Footroom clearance underneath table is at least 19 inches?
____ Table legs are 30 inches apart?
____ Pedestal-base tables have low tapered bases and a minimum diameter of 42 inches?

Self-Service Areas

____ Food service lines are 36 inches to 42 inches wide?
____ What is counter top height?
____ There is clear floor space between the end of the food service line and counter for proper turning?
____ Salad bar provides 36-inch clearance on all sides?
____ Salad bar provides area to temporarily set plate?
____ Salad bar height is 34 inches?
____ Tray slide underneath is 29 inches in height?

___ Condiments accessible?
___ Tilted mirror provided above food display?
___ Utensils and napkins are accessible?
___ Will food be cut upon request?
___ Menus are available in braille?
___ Will servers read menus on request?

Guest Rooms
Entrance

___ At least one main entrance is accessible to wheelchairs?
___ Door hardware is standard or lever handle?
___ Doorways have clear openings at least 32 inches wide?
___ Door openings are no more than 5 lbf?
___ Speed of door closure is at least 3 seconds?
___ Doorsill has no sharp inclines and/or abrupt changes in levels?
___ What is door hardware height?
___ Card entry system allows 10-second entry?
___ Electronic warning system is available from hotel management to use with the hotel smoke alarm, telephone, and clock radio?
___ Electronic door knock alert is available?
___ Audible smoke detector is available?
___ Visual smoke detector is available?

Room

___ Maneuvering space at door is 5 square feet?
___ Pull-side has 1½ feet clear space?
___ Door opening aid is available?
___ Room has maneuvering space throughout?
___ If floor surface is carpet, type and height of pile?
___ Windows:
   ___ Height? Type? Openable?
   ___ Latch location and height?
___ Light switch height is approximately 48 inches from floor?
   ___ Jacks are provided for portable computers and/or fax machines?
   ___ Computer/fax jack is neither higher than 48 inches from floor nor lower than 12 inches from floor?
___ AC/heat control is accessible?
___ Telephone:
   ___ Within reach, with volume control (amplifier) or TDD available upon request?
   ___ Analog lines available for TDD?
   ___ Hearing aid compatible (Magnetic)?
   ___ Audible message signal?
   ___ Visual message signal?

____ Mirror accessibility:
     ____ Height from floor?
     ____ Dimension of mirror?
     ____ Location?
____ Television:
     ____ Remote control for TV?
     ____ Both broadcast TV and pay TV?
     ____ Can be differentiated without sight?
     ____ Closed captioned viewing available? Telecaption decoder available? Cable ready with remote?
     ____ Folio review and checkout available?
     ____ In-room shopping available?
     ____ Room service ordering available?
     ____ Guest message broadcast channel?
     ____ Area information channel?
____ Beds:
     ____ Single? Double?
     ____ Bed height 18 to 20 inches from floor?
     ____ Clearance alongside bed at least 3 feet?
     ____ Approachable from either side?
     ____ Kickspace below box spring 3 inches high and 3 inches deep?
     ____ Adjustable bedside extension mirror?
     ____ Transfer board available?
____ Bedside tables:
     ____ Table height 3 inches above mattress?
     ____ Reading lights?
     ____ Clock radio?
     ____ Light switches accessible within a 24-inch reach from the bed?
     ____ Remote controls for TV and/or radio accessible from bed?
     ____ Emergency warning signals audible/visible from bed?
____ Other furniture:
     ____ Dresser hardware?
         ____ Staple handle?
         ____ Half-inch clearance?
     ____ Desk:
         ____ 29 inches high by 30 inches wide?
         ____ At least 19 inches of foot room clearance?
     ____ Reacher aid available?
____ Hallways have 36-inch width or more?
____ Closet:
     ____ Closet type? How opens?
     ____ Closet bar height?
     ____ Closet shelf height?
     ____ Closet safe?
         ____ Where placed? Accessible?

____ Bathroom:
     ____ Inside door width 32 inches minimum?
     ____ Lever handle or push/pull type of door hardware?
     ____ Turning radius 5 feet × 5 feet for wheelchairs?
     ____ Space to maneuver to toilet?
     ____ Approach to toilet?
         ____ Diagonally? Straight on? Back up to one side?
         ____ Portable raised toilet seat with adjustable arm support?
     ____ Toilet seat and paper dispenser height 17 to 19 inches?
     ____ Grab bars 33 to 36 inches high?
         ____ Parallel to the floor?
         ____ 1.5 inches between rail and wall?
         ____ Sturdy?
     ____ Wash basin/sink:
         ____ Clearance of 29 inches underneath?
         ____ No higher than 34 inches at top?
         ____ Faucet hardware: Standard? Lever? Single lever?
         ____ Plumbing is insulated?
     ____ Mirror no higher than 40 inches from floor?
     ____ Towel rack accessible?
         ____ No higher than 54 inches from the floor and to the side of the wash basin/sink?
     ____ Light switches, electric outlets:
         ____ No higher than 48 inches if forward reach?
         ____ No higher than 54 inches if side reach?
     ____ Bathtub or shower:
         ____ Lined with nonskid surface?
         ____ Grab bars placed along the walls?
         ____ Handheld shower?
         ____ Tub height?

F i v e

Reservations

The reservations department is a key department of any guest establishment. A prospective guest calling for information or to make a reservation often forms his or her first impression of the establishment from the response, inflection of voice, and general conversational tone of the reservations agent. This initial contact with the establishment will suggest how responsive the establishment is to guest services, and how much importance it attaches to guest courtesy within the establishment. The tone of the reservations agent, his or her voice modulation, enunciation of words, and speed of speech convey a "feeling" to those on the other end of the telephone line. You want that feeling to convey, "They care about me [the prospective guest] and want to make my stay there enjoyable."

❖ CHOOSING A RESERVATIONS AGENT

Reservations agents should be happy, upbeat people who enjoy being with, talking to, and helping other people. Ideally, they should be bilingual or trilingual.

In hiring people, the first and most important rule is to seek applicants with a great attitude—people who enjoy what they do and have done in the past. Hire prospects who believe in your establishment's quality, atmosphere, service, ambience, and food. When hiring, ask questions designed to ferret out people who take pride in conducting themselves in a professional manner and believe their work centers around serving the guests, not taking coffee breaks to catch up on employee gossip. Ask prospective employees these questions:

- ◆ What did you enjoy most about your last position?
- ◆ Why do you want to work here?

♦ What do you know about our company?
♦ What job goals are important to you?
♦ What are your career goals for the next 5 years?
♦ How would you introduce yourself to a guest?

Describe several scenarios and ask the prospective employee how he or she would handle the situation. Compare the applicant's responses with company goals and evaluate whether the responses suggest the person will take the position seriously, will want to advance in your company, will take pride in doing a job well and will accept personal responsibility for taking care of guests' needs, problems, or complaints. Will this person resolve the situation within the limits of what an employee is empowered to do, (for example, offer a complimentary dessert, offer a complimentary additional night's stay, offer a complimentary continental breakfast).

❖ RESERVATIONS PROCEDURES

Company policies and procedures must be established and enforced, and updated and/or revised when necessary. When policies are revised, it is imperative to hold staff meetings, first to discuss the events that have led to the need for the revisions (e.g., increase in non-English-speaking guests); then to review the new policies (e.g., company reimbursement for second-language lessons, differential pay for employees who speak additional languages). In addition to staff meetings, formal notices of the changes must be posted in conspicuous places, such as the human resources office, employee cafeteria and lounge, and employee entrance and exit areas. An excellent way to disseminate employee information is to enclose a leaflet, brochure, or statement about the new policy with the paycheck before that policy goes into effect.

Prompt Telephone Service

A common policy in reservation departments is one stating that the telephone lines in the reservations department are reserved for business use only. Corollaries to this basic policy address personal calls and incoming messages for employees. The policy regarding reservations phone lines might state:

> Personal calls are not permitted on reservation office telephone lines, incoming or outgoing. Personal phone calls should be made from pay telephones located in the staff lounge and near the

employee entrance. Incoming phone calls to employees are limited to emergency messages only. Such messages will be relayed to the employee's immediate supervisor who will relay the message to the employee.

The employee handbook should contain a paragraph or two similar to the following:

> Messages will not be taken for employees unless it is an emergency. If there is an emergency, have the person calling state that it is an emergency and the message must be taken to you. A message will be taken and given to your supervisor, who will then give you the message.

Basic phone courtesy dictates that the phone be answered promptly and courteously. When there are limited phone lines and many incoming calls, a potential problem, or a challenge, is created.

A guest can be asked, "May I please put you on hold for a moment while another call is answered?" If the guest agrees and is put on hold, the receptionist must get back to the first caller in as short a period of time as possible.

Some establishments install a system that will automatically give the following message:

> All phone lines are currently busy, but if you (the caller) remain on the line, your call will be answered in the order it was received.

Other establishments have a message stating, "The next available agent will take your call; please remain on the line," or words to that effect.

After the message, music or a prerecorded message detailing the highlights of the establishment might play continuously until an agent answers the call.

Answering the Call: What to Know

When answering calls, the receptionist must be courteous, friendly, and knowledgeable. People may be calling for a variety of reasons: to obtain information, to make a reservation, to order a birthday cake, or to ask for all these services. The reservations agent should:

◆ Answer courteously.
◆ Use a first name in identifying him or herself to callers.
◆ Be friendly and polite, yet businesslike and professional.

RESERVATIONS AGENT'S CHECKLIST

A Basic Checklist of Items Reservations Agents
Need to Know

_____ Locations and days and hours of operation of restaurants,
gift shops, beauty salons, health fitness facilities, busi-
ness centers, and day-care centers at the establishment.
_____ Directions to get there by public or private transporta-
tion (car, taxi, bus, subway).
_____ Dress code requirements for restaurants, if any.
_____ Specialties of the house for each food outlet or restaurant
within the hotel and in surrounding area.
_____ Scheduled special or private parties (e.g., bridal shower
for Jane Jones is next Sunday at 2 P.M. in the Red Room).
_____ Area attractions and directions to get there.
_____ Key people and their beeper numbers:
   _____ Resident manager.
   _____ Restaurant managers.
   _____ Banquet managers and private dining managers.
   _____ Sales personnel for private dining, conventions.
   _____ Group sales manager.
   _____ Personnel or human resources manager.
   _____ General manager.
_____ Access for the disabled information.
_____ Meal periods for each food outlet (e.g., lunch, 11:30 A.M.–
3 P.M.; dinner, 4:30 P.M.–11:30 P.M.; seven days a week).
_____ Forms of payment accepted (e.g., cash, MasterCard/Visa,
American Express).
_____ Reservations policy (e.g., 1 month in advance for the
gourmet restaurant; no reservations at the family restau-
rant; up to 2 years in advance for private banquets).
_____ Holiday reservations policy (e.g., Mother's Day reserva-
tions taken 2 months in advance).

The agent must listen to the caller carefully, quickly ascertain
what information is needed, and disseminate that information. A
basic checklist of items reservations agents need to know is shown
above.

Dealing with Job Applicants

Reservations agents may receive many employment inquiries in
addition to calls for meal reservations. Reservations agents need

to know where to obtain an application, when applications are accepted (what days of the week and what hours on those days), when interviews are granted, whether there are any current openings, and whom to contact after completing an application.

Menu Information

To answer many of the food and beverage questions, reservations agents need access to all menus for each food outlet:

♦ Breakfast
♦ Brunch
♦ Lunch
♦ Pretheater
♦ Dinner
♦ Dessert
♦ All-night or late-night snacks
♦ Room service (in hotels)
♦ Wine lists
♦ Cordial/aperitif lists.

Agents should also have answers to questions such as the following: What types of beer are available? Is the fish fresh or frozen or a combination? Are the steaks fresh or frozen? Aged?

Making a Restaurant Reservation

The agent taking a restaurant reservation must note basic information plus any special considerations in the reservations book or on the computer, if one is used. The agent's initials or code should be entered on the reservation as well as the current date.

The following information should be requested:

♦ Name of party, including first name.
♦ Number in party.
♦ Phone numbers where contact person can be reached (home; office; and hotel) and/or a local phone number where they can be reached when in town if calling from out of state.
♦ Date and time of reservation.
♦ Special requests, such as wheelchair, high chair, birthday or anniversary cake, chilled champagne at table on arrival.
♦ Room of reservation if several dining rooms are available.
♦ Method of prepayment, down payment, or deposit. If by credit card, be sure to obtain credit card type, credit card number and expiration date, and obtain an authorization code.

♦ VIP status. Is this person VIP (very important person) at this establishment?

To verify accuracy, the reservation agent must repeat all information gathered from a caller, including the spelling of the guest's name. The agent must also state to the guest the cancellation policy, the holding time for tables, and the confirmation policy, as well as the confirmation number, if the establishment uses that procedure (e.g., reservations must be confirmed that day or they may not be honored).

For instance, the agent might say:

Mrs. Smith, is that spelled S-m-i-t-h? [If yes, the person continues.] Your reservation for six people in the Blue Room for Saturday, October 24 at 8:15 P.M. is confirmed. We will be expecting one person in a wheelchair. Your table will be held for 15 minutes, that is, until 8:30 P.M., at which time it may be released to a waiting party. If you will be delayed for any reason, please call. If you need to cancel your reservation, we ask that you call at least 24 hours in advance of your reservation. Do you have any further questions [or, May I be of further service]?

If any part of the information is incorrect, it should be corrected first, then the conversation should progress as described.

If a guest has requested a particular time or room that is not available, be sure to suggest an alternative. If the person wanted a dinner reservation in the Green Room at 8:00 P.M. on Saturday night, but the Green Room is booked for a private party, suggest the Blue Room. Or, if the Green Room is booked with other reservations at 8:00 P.M., suggest dining in the Green Room at 6:30 P.M. or 9:30 P.M. If possible, offer the guest several choices.

VIP Status

Often regular patrons, family members of the owners, celebrities who are in town, politicians, and others in the public eye will be granted VIP status. Generally, the managers and/or owners designate these special guests. The VIPs are generally seated at the best tables in the best rooms, and the chef will pay extra attention to the plates going out into the dining room. A complimentary hors d'oeuvre, first course, or dessert may be brought to the table.

The VIP status is usually accorded to guests who are likely to attract additional business. Many people may come to the establishment hoping to see a celebrity and to dine at the next table, or at least in the same room. The extra cost of complimentary dishes is written off as an advertising or marketing cost of doing business in the hope of attracting future guests.

Cake Orders

If the restaurant provides cakes for birthdays, anniversaries, or other special occasions, the reservation agents need to have a cake order sheet specifying the following information:

- Size (6-, 8-, 10-, or 12-inch round, half- or full-sheet cake).
- Cake type (chocolate, vanilla, fat-free angel food cake, etc.).
- Type of frosting.
- Inscription.
- Prices.
- Cancellation policy.
- Deposit or prepayment amount (if credit card, credit card number, expiration date, authorization code).

The cake order sheet should be written in triplicate. The top sheet goes to the pastry chef/bakery department; the middle copy remains in the reservations office, and the third copy is for the restaurant manager, who will need to know there is a cake for that party.

In addition to the above information, the cake order must contain all the standard reservations information, including a phone number where the contact person can be reached (in case it is a surprise party and the pastry chef has any questions).

The confirmation for a cake order might be as follows:

> Mrs. Smith, in addition, you are ordering a 10-inch angel food cake with fresh raspberry topping. There will be a banner above the cake saying, "Happy 80th Birthday, Harry." We understand that this is a surprise party and only you should be called regarding this reservation or this cake. The cake is $25. If you need to cancel the cake, we must be notified 48 hours in advance or you will be charged for the cake. A 50 percent deposit is required no later than one week prior to the reservation date. Would you care to pay that now by credit card?

The reservations agents should have a guide showing approximately how many portions are in each size cake. Thus, if a guest wants a cake for 6, 8, or 12 people, the agent will be able to assist in choosing the proper size.

Wedding cakes may have special stipulations, such as requiring a minimum notice of 1 week (preferably 2 weeks). There also may be an additional charge for decorating such cakes, a longer cancellation time requirement, and a larger deposit.

SAMPLE CAKE SIZE GUIDE		
Size	*Number of Servings*	*Price*
6-inch round	1–8	$15
8-inch round	8–12	20
10-inch round	12–16	25
12-inch round	16–20	35
half sheet	20–40	45
full sheet	40–80	55

Balloons and Other Giveaways

On special occasions, such as a birthday, anniversary, or baby shower, balloons add to the festive atmosphere. If the establishment has a balloon policy, the reservations agent should mention it:

> Mrs. Smith, we have complimentary balloons that add a festive air to the table. Shall we have some balloons at the table on your arrival?

Preordering Special or Additional Items

A special occasion is also a perfect opportunity to have a bottle of champagne or nonalcoholic sparkling juice waiting at the tableside, or one of each. Having them waiting gets the party off to a great start—no one has to wonder what to order and no time is lost while the server checks to see if the chosen bottle and vintage are in stock, gathers up the ice bucket, fills it with ice and water, gathers the champagne glasses, and brings all to the table.

The reservation agent might suggest:

> Mrs. Smith, an 80th birthday party sounds very special. In addition to the birthday cake at the conclusion of the meal and the balloons that we'll be very happy to place on the table for your arrival, would you like to add a bit of sparkle to the occasion by having a bottle of champagne or nonalcoholic sparkling juice at the table so you can make a toast when the guest of honor arrives?

If Mrs. Smith agrees, the type of champagne and/or sparkling juice requested must be noted.

For Mrs. Smith's reservation for October 24, there would be several special considerations noted on the reservations sheet or in the reservations computer:

- ◆ One person arriving in wheelchair.
- ◆ Angel food w/raspberry birthday cake: SURPRISE.
- ◆ Balloons on table.
- ◆ Dom Perignon chilled at table.
- ◆ Sparkling apple juice chilled at table.

❖ RESERVATIONS ERRORS

It's not a pleasant experience when people show up at 8:00 P.M. on a busy Saturday night insisting they had a confirmed reservation, and the maître d'hôtel or other front desk personnel cannot find it.

Good advance planning will help differentiate those who do not have a reservation from those whose reservations truly did get lost, although there is virtually no difference in how both are handled—all guests who want to dine should be accommodated as rapidly as possible.

One method to identify "phony reservation patrons" is to stagger reservations at odd times such as 8:10 P.M. or 8:40 P.M., but not to have an 8:00 P.M. or 8:30 P.M. seating. Some establishments create such seating times to identify patrons who abuse the reservations policies. If a person becomes a "repeat offender" the maître d' might call the guest the following day and gently suggest something to this effect:

> We love to accommodate your dinner (lunch) parties, so in the future, to avoid the kind of confusion we had last evening, could you please make and confirm your reservations with me personally?

If the restaurant has a policy not to make reservations on the hour or half hour, a person who arrives claiming to have made a confirmed 8:00 P.M. reservation obviously is in error. Perhaps the party had no reservation at all but is hoping for a table, or had an 8:15 P.M. reservation and forgot the time, or had an 8:00 P.M. reservation at a restaurant with a similar name and arrived at your restaurant by mistake.

Don't embarrass the potential customers, and don't send them away. If they are sent away once, they might never return, even if they didn't have a reservation and are just trying to "squeeze in." If a table is available, seat the party. If not, assume that it is the house's mistake, whether it is or not, and assure the party, "We must have made an error; we'll arrange a table for you as soon as possible." Then do so; that is, follow up on exactly what you've promised and arrange a table for these guests as soon as one becomes available. How they are treated will determine whether or not they, and their friends and associates, return. If they are out-of-town tourists and never do return, how you treated them will definitely influence their memory of their visit to your city and which establishments they recommend to friends and relatives who might be coming to town in the future.

Often it helps to offer a complimentary cocktail to "missing reservation" guests as it may be 30 minutes or so before they can be seated. This accomplishes several goals: The guests are not embarrassed, they are made to feel special and wanted, and the staff gain time in locating a table. The party frequently will thank you profusely for accommodating them and making their meal special.

If a person arrives claiming to have made a confirmed reservation for 8:10 P.M. (a time the restaurant uses), it may very well be legitimate (or the party's been there before). Ask when the reservation was made and check the reservations sheets (or the computer). Perhaps the name was misspelled or the date was erroneous—it's amazing how many people will show up for their reservation a week early or a week late! In any event, seat the guests as soon as possible. Make them feel welcome and ensure they are comfortable while waiting.

The reservations agents can help greatly by identifying themselves and using their first names (or a "working" first name) when they speak with callers. The caller might remember whom they spoke to, lending credibility to their claim of having a reservation.

❖ ADVANCE RESERVATIONS

Advance reservations are those for a date beyond the reservations book dates or computer span of dates for which you are currently accepting reservations. For instance, it might be a reservation for a birthday dinner party for eight people on September 23, but it's now May 3 and the restaurant only books reservations 1 month in advance. These are often mail requests and should be honored as soon as feasible. If the request is within the current reservations period, "book" it and call the party to confirm and arrange any details not noted in the mail request.

If the reservation request is outside the current "booking" period, hold the reservation request until it can be honored within the guidelines of the house reservations policy. For example, a restaurant has a 1-month-in-advance reservations policy (2 months for major holidays). The restaurant receives a request on January 3 for lunch for four people on Independence Day, July 4, at 2:00 P.M. in the Red Room. That request cannot be honored when received. The request should be filed under May 4 in a chronological "tickler" file, that date being two months prior to July 4. Every day, before the reservations office opens the phone lines for reservations and customer assistance, the advance requests corresponding to the current date should be removed from the tickler file and entered as reservations. Then the reservations department can fill in the remainder of the available reservations slots for that day as people call. Thus, the mailed reservation request for July 4 would be pulled from tickler file and entered into the reservations book or onto the computer on May 4. Then the party either will be mailed a confirmation number and asked to call if there are any special requests, or will be called, given the confirmation number, and asked if there are any special requests.

❖ PREPAYMENTS

Prepayments are often accepted when a person wants to "treat," or pay the bill for a group or party, such as a child's birthday party, or a business luncheon or dinner. Ideally the person will visit the establishment, prepay the bill, and review the following details of the meal:

1. Should there be balloons at the table?
2. Should any alcoholic beverages be offered? Will the guests be of legal age to consume alcoholic beverages? If it is a business meeting, should alcoholic beverages be offered, or does the company have a no-alcohol policy and want alternative beverages to be offered?
3. If there a special menu? If it's a birthday celebration, the birthday person may have a favorite dish; perhaps there will be vegetarians in the group.
4. Should a cake be ordered? If so, what size, what type, what inscription?
5. How much is the gratuity? Often, for parties larger than eight persons, a gratuity will automatically be added. The percentage,

which varies depending on the establishment, typically ranges between 15 and 23 percent of the total check.

If the menu and beverages are predetermined and the gratuity percentage is settled, the bill can be prepaid in full. Otherwise a partial payment or deposit can be paid when the reservation is made, with the remainder billed to the payor at the conclusion of the meal. Another alternative, often used for business meetings, is to review expected charges with the payor. Then ask the payor to sign a blank credit card voucher and have it authorized for the "guesstimated" amount. At the conclusion of the meal, the check is totaled, the predetermined gratuity is added, and the credit card voucher is processed. The payor is mailed a copy of the complete bill, plus the payor's copy of the credit card voucher.

❖ HOLIDAY MENUS AND BOOKING QUESTIONS

Special preset menus are often offered for holidays, such as Thanksgiving Day, Mother's Day, Christmas Day, Easter Sunday, and New Year's Eve. There are two common scenarios: *prix fixe* and *tiered pricing*.

Prix Fixe

One price for all, or one price for adults and one price for children (usually between the ages of 4 and 12 years).

The meal is usually a buffet or a limited menu meal. With high volume, the buffet is easiest for both kitchen preparation and traffic handling. A limited menu with one price per person might offer two or three appetizers, one salad with a choice of two or three dressings, two or three entrées (perhaps one chicken, one beef, and one fish), and two or three desserts, as shown in the sample prix fixe menu.

Tiered Pricing

With tiered pricing, several choices of appetizers and entrees are available, each with a different price. A modified version of tiered pricing is having a price supplement for certain entrees such as steak or lobster.

PRIX FIXE HOLIDAY MENU

Melon in Season
Cup of Homemade Mushroom Soup
Shrimp Cocktail Salad

*   *   *

Fresh Salad Greens with Choice of Dressing
(French, Italian, Blue Cheese)

*   *   *

Roast Chicken Breast on Bed of Spinach
Roast Filet of Beef with Roast Potatoes and Vegetables
Baked Sole with Garden Vegetable Medley

*   *   *

Angel Food Cake with Fresh Strawberries
Decadent Chocolate Flourless Cake
Sorbet or Ice Cream

*   *   *

Coffee, Tea, or Soda

$19.95 per adult   Taxes and gratuity
$ 9.95 per child under 12   not included

Tiered menus are especially useful when groups are scheduled for holiday meals. The establishment might have three or four complete dinners, each at a different price. Organizers of the group preselect the meal and notify the establishment of the choice and any required deviations, for example, how many vegetable platters they will need; one guest who can eat only soft foods. The tiered sample menu shows this kind of holiday menu.

For certain holidays such as Thanksgiving, there may be a traditional primary menu (roast turkey, gravy, mashed potatoes, candied yams, broccoli, green beans, pecan pie, pumpkin pie, and apple pie à la mode) and a secondary entree option (e.g., baked fillet of sole), for those who don't wish to order turkey.

Guests who call to make holiday reservations should be advised of all holiday menus and options, including supplemental charges. Each menu item should be read to guests calling for reservations, so they will know exactly what the selections will be and

HOLIDAY MENU WITH SUPPLEMENTAL AND
TIERED PRICING

Onion Soup Gratinee
Fresh Fruit Cup
Sevruga Caviar Service ($25 supplemental charge)

*   *   *

Fresh Salad Greens with Choice of Dressing
French, Thousand Island, or
Gorgonzola Blue Cheese ($1.00 supplemental charge)

*   *   *

Roast Stuffed Chicken Breast
Fresh Sole with Garden Vegetable Medley
Roast Filet of Beef with Roasted Potatoes
$15.95

Poached Salmon Steak with Artichoke Hearts
Shrimp and Lobster Claw Garnish
$22.95

Medallions of Filet Mignon with
Wild Mushroom and Cognac Sauce
$24.95

Roast Muscovy Duck Breast with Lingonberry Sauce,
Quail and Pheasant Terrine
$26.95

*   *   *

Fresh Berries in Season
Chocolate Mousse
Sorbet
Bananas Foster ($2.50 supplemental charge)

Dinner includes choice of appetizer, salad, entree, dessert,
and coffee, tea, or soda.

Taxes and gratuity not included.

the price of each meal. In other words, there should be no surprises when they arrive!

Special seating arrangements may be needed. Often, a restaurant will have a "normal-day" seating chart and a "holiday" seating chart with a larger number of tables that seat six or more. On ordinary days and weekends, restaurants utilize many deuces (two-tops) and four-tops for business meetings of 2, 3, or 4 persons, off-site interviews, friends meeting for lunch, and couples. On holidays, however, typically whole families and extended families—mom and dad, grandma and grandpa, the kids, an aunt and uncle, a cousin who happens to be in town, and a friend—may want a table together. Tables for 6, 8, 10, or 12 are much in demand during holiday meal periods, and tables for 2 (especially) and 4 have limited use, although tables for all size parties are needed to optimize seating.

A holiday seating chart should be created if there is not one. The majority of tables on the chart should be for 6 or more people. Once the property has a "holiday" history, the number of 6-tops, 8-tops, 10-tops, and 12-tops needed during each holiday meal period can be estimated more accurately. Usually, the property accomplishes the change in seating by combining the 2- and 4-tops to create 6- and 8-tops.

Ideally, rounds are used for large parties (6 or more) as they are better for conversation; however, very few properties have the storage space for the smaller deuces and 4-tops and for the larger tables during non-holiday periods. Some properties use tables with four hinged leaves. When these four leaves are folded under the tabletop it can be used as a square 4-top; opening the leaves creates a round table suitable for 6 people.

Other properties have large wooden 8-top, 10-top, and 12-top rounds cut to their specifications with guide pieces on the bottoms. These rounds fit atop square tables, changing them into 8-tops or larger. The guide pieces prevent the large round from sliding off the square table, which serves as the base.

For holiday bookings, reservationists must be sure to use a holiday reservations chart and not the normal-day chart. If an agent begins booking holiday reservations with a normal-day reservations chart, the few large tables will be booked almost immediately, and business may be turned away because it mistakenly appears as though there are no remaining large tables. Likewise, small parties may be turned away if a holiday seating chart is used for an ordinary day.

The advent of computer reservations booking has eliminated most of the manual error from reservations booking. A computerized reservations system usually has holidays programmed into it so the computer knows when a holiday occurs and immediately switches to the holiday reservations booking chart.

❖ GUEST COMPLAINTS

Written Responses

Sooner or later, an unhappy, disgruntled guest, former guest, or potential guest will call or write. The room was too small; the air conditioning or heat didn't work; the TV was broken. Or, the restaurant meal was awful; the coffee was cold; the service was terrible; the entrée didn't arrive for over an hour and dessert never arrived.

Every establishment should have a specific, written "customer complaint" or "angry guest" policy and should communicate it to all customer service representatives, managers, and other "key players" including reservations and PBX agents. The main points should be posted near the customer service agents' workstations and provided to everyone involved in answering customer complaints. These procedures should be followed scrupulously to ensure that guests' complaints are not only addressed but resolved in a satisfactory manner.

Some establishments ask one or two of their customer service representatives, reservations agents, or reservations supervisors to draft preliminary responses to customer complaint letters. These letters are then reviewed, edited, retyped onto letterhead stationery, and signed by a manager.

Written complaints should be forwarded immediately to the appropriate department manager. Every written complaint should be answered in a timely manner, at least within a month. The letter should address the guest's complaint, apologize for any inappropriate actions or nonactions of the establishment, suggest that whatever happened is not the norm nor usual policy of the establishment, and offer compensation or complimentary item(s) or service(s), or superlative service on the guest's next visit.

If a server spills something on a guest, the establishment usually will pay for the dry cleaning of the garment(s). An accident report should have been completed at the time, noting the person(s) affected by the spillage, the item(s) damaged, and the extent of the damage to each item. The dry cleaning bills, when received, should be cross-referenced against the accident report to ensure that the establishment pays for the dry cleaning of only those articles damaged during that incident. Otherwise, a guest might submit and be reimbursed for a month's worth of dry cleaning bills.

If the complaint is about sloppy service in general, management might enclose a gift certificate for the guest's next visit. The amount of the certificate would depend on the price scale of the restaurant and the seriousness of the perceived or real complaint. A restaurant might present a decorative "Appetizer/Dessert Voucher"; a hotel or resort property might implement a "Weekend Stay Voucher." The voucher should be designed by a commercial artist who can incorporate the motif or logo of the restaurant or hotel, and the vouchers should be printed in four-color process on glossy or quality bond paper.

This colorful, customized voucher can be sent with the guest's name imprinted on it and an expiration date (usually 1 year). The expiration date is a subtle form of marketing. Having an expiration date is an attempt to influence the guest to return soon so the memory of the troubling incident can be replaced by a positive experience, which in turn would transform their formerly negative word-of-mouth advertising into positive comments. Nevertheless, whether or not the guest ever uses the voucher or gift certificate, the establishment's effort to resolve the problem is beneficial and will usually mitigate any negative customer feelings about the original incident.

Verbal Responses

Some guest complaints will be verbal, either from someone appearing in person or calling on the telephone. Reservations agents, as well as customer service representatives and managers, must be trained to handle these complaints. The establishment's complaint policy should be documented and a copy given to each reservations agent receiving new-hire training or cross-training for the reservations department. The person must practice handling typical problems of guests, including role playing if possible, prior to dealing with live telephone complaints.

Procedures for Handling Complaints

A highly recommended approach to handling guest complaints follows. Key points include being empathetic to the guest, showing understanding of the guest's concern, and arriving at a solution. Once the parties agree on a solution, the employee must try to implement it immediately and assure the guest that whatever was promised will happen in a timely manner. It is essential to

STEPS IN HANDLING GUEST COMPLAINTS

1. Personalize the situation: Give your first name and use the guest's name.
2. Listen actively and intently.
3. Obtain the facts.
4. Be understanding; show care and empathy.
5. Don't react personally to the situation.
6. Generate a solution to the problem.
7. Act immediately. Do what you said you would do, and be professional at all times.
8. Follow up on the situation.

follow through on such promises. The solution should be within reason and should utilize good business sense.

Personalize the Situation

Find out the guest's name and use it often during the conversation. Give the guest your first name and repeat it so the guest will know whom he or she is speaking to and can follow up with you.

Some establishments allow customer service representatives to adopt a "phone name" to ensure their privacy. A phone name is a first name used when speaking with customers that is not necessarily the person's given or legal name. As long as the employee consistently uses the same name whenever working, it doesn't matter whether the name is real or invented.

Listen Actively and Intently

Let customers say what they called to say. If they need to "sound off," allow them to do so, uninterrupted. If a guest continues for more than a minute, murmur an intermittent, "Yes, I understand," or "Uh, huh," to demonstrate that you are still listening.

Usually customers will feel better and be more receptive to ideas after they've had a chance to "state their case." While you must allow the guest this freedom, it is important to maintain control of the conversation. Keep the guest focused on the complaint at hand, and do not allow the conversation to turn into a general gripe session or recital of the guest's life history.

Be sure to speak in a professional manner when you interject remarks to maintain control. You might refocus a wandering complaint by reminding the person, "As you were saying a moment ago, in regard to [the complaint] . . ."

Obtain the Facts

Encourage customers who are upset to describe what upset them, what happened, in detail. Take the time to obtain all the details that they remember, and take notes of what the guest is saying.

Ask open-ended questions in a professional, nonthreatening manner. Summarize the problem. Make sure you understand all the facts of the situation as related by the guest.

Be sure to obtain the guest's name, including its correct spelling, as well as the person's phone number and address. At a later date, this information may be needed to send a letter of inquiry or apology, along with a gift certificate or voucher of some sort.

Repeat what the guest has said; summarize the main points, using the guest's words. Be efficient yet courteous, gentle yet effective.

Be Understanding

Listen attentively and as mentioned earlier, say, "Yes, I understand" intermittently to demonstrate empathy. Show that you are listening.

Imagine yourself in the guest's place. How would you feel if you experienced the situation, as related by the guest? In that frame of mind, offer a sincere apology and assure the person that you want to be helpful.

Try to let the caller hear the "smile" in your voice. Your job is to make the caller a satisfied customer before breaking the telephone connection.

Don't React Personally

While you want to personalize the conversation by addressing the guest by name and having the guest address you by name, you do not want to react emotionally to anything the guest says.

Reacting emotionally tends to escalate or intensify the situation. Your goal, instead, is to defuse the situation, to resolve any differences, and to ensure a satisfied customer who will become a repeat customer.

When an irate guest continuously states that "you" did this or that, or didn't do this or that, don't take it personally. Repeat, "I'm sorry that you feel . . ." Describe whatever the complaint revolves around—service was too slow, the room was too hot/cold, but do not admit liability by agreeing to the guest's unsubstantiated complaint. Reiterate, "I understand that you believe . . ."

Generate a Solution

Offer alternatives, at least two, but suggest only choices you are authorized to propose. A solution does not necessarily admit that what the guest stated did happen. Rather, it suggests that as a valued customer the establishment would like to offer something to let the guest know 1) they are a valued customer, 2) that the establishment is sorry about the guest's feeling that (whatever the complaint was), and 3) please accept this (solution) as our way of saying, "We're sorry for any inconvenience you may have experienced."

Ask how the guest would like to see the problem resolved. Zero in on a solution. This is most important. Sometimes all the person wants is an apology; sometimes a complimentary dessert or dinner will suffice; sometimes a complimentary room for one night will be acceptable. Determine exactly what the guest wants, and then explain what you *can* do to solve the problem.

Thank the guest for bringing the matter to your attention so that you can take care of it. Make the guest feel that it is a privilege for you to do so and that resolving the concern satisfactorily is the most important item on your agenda, that you will do whatever it takes to see it through to an acceptable solution.

Decide on a course of action that is mutually acceptable. If you can satisfy the guest's request, say so promptly and give a date when it will happen. For example, if service in the restaurant was exceptionally slow causing a couple to miss their theater play and they will accept a "dinner for two" voucher, state that you will send the voucher today and they should receive it within the week. Further state that if they do not receive it within the week, they should call you, but, regardless, you will call them 2 weeks from today to verify that they have received the voucher and are satisfied with the solution.

If you cannot solve the problem to the guest's complete satisfaction, state exactly what you can do and stress that you will follow through. Then do whatever you promised to do. The action might be as logical as submitting a report to the manager of the department to ask for whatever the guest has requested, and calling the guest back in two days to advise of the next step.

Should you mention to the guest that the next step is to have a manager assess the situation, give the guest the name of the manager who will be handling the situation.

Act Immediately

Relate again to the guest the agreed-on solution. Give the guest the timeframe to enact the solution. State your first name again to the guest and ask the guest to call at any time to monitor the situation.

Act immediately. Take personal responsibility for whatever was promised to the guest, and ensure that it happens in a timely manner, preferably within 1 month of the initial incident.

Follow Up

Follow up by calling the guest within 2 weeks after the initial phone call even if the situation has not been totally resolved. Follow up with the guest at least once every 2 weeks thereafter until the situation has been resolved. Use a tickler file to track who needs to be called on which dates.

At the conclusion of the situation, when whatever has been promised to the guest has happened, follow up again—call to ensure that the guest is satisfied with the solution. Extend an invitation to such guests to return, to try your establishment again. Make them feel welcome. Let them know that you value their patronage and that their thoughts regarding your establishment are important—that's why you acted on their complaint.

Over 90 percent of customers with a complaint will return to your establishment again if you resolve their complaint quickly and satisfactorily and if you make them feel they are valued customers.

When Guest Requests Can't Be Accommodated

Sometimes, a guest cannot have what he or she wants. A sample scenario might be the guest who writes after a 5-day stay stating that his bed was uncomfortable and he wants a complimentary weekend to compensate for this. The records are checked, and this guest never lodged a complaint during his stay.

When responding to such guests, use the following guidelines:

◆ *Begin on a Friendly Note.* Thank them for their patronage and tell them you hope they enjoyed their stay. State that you're sorry that . . . [whatever they complained about].
◆ *Tell Them Why You Can't Fulfill Their Request.* Be specific about their complaint, be logical, and be businesslike yet

friendly. Frankly state why you can't accommodate their wish. Use an explanation that will bring them to the same logical conclusion. When they finish reading your letter, they should understand why their demand was unreasonable and should agree with your conclusion.

◆ *Finish on a Friendly Note.* Thank them for bringing this matter to your attention. Tell them that you'd love to have them visit your establishment again, that you're looking forward to hearing from them again, and that you hope they have a wonderful time on their next visit. Let them know you appreciate their interest in your establishment.

◆ *House Policy Permitting, Give a Gift.* If house policy permits, enclose a gift of nominal value (e.g., a dessert voucher). This sign of goodwill indicates a compromise, creating a "win-win" situation. You didn't give the guests what they asked for, but you did give them something that they didn't have prior to writing to you: It resulted from your attention to their complaint. They are acknowledged.

Various Guest Complainer Personalities

It may seem as if you have the same guest complainers over and over again; they just wear different clothes on different occasions, seem a little taller one night than another, or a bit heavier, or somewhat younger, or older. Otherwise, they could be the same person, complaining over and over again.

There are patterns to complainers. The following sections describe a few of the most common types; can you relate to any of them?

Troublemaker

The troublemaker is the guest who's never satisfied with anything, who cannot be pleased no matter what you do. Nothing is ever right. He or she will criticize everything. The bed's too hard; the bed's too soft. The service in the restaurant is too rushed; the service is too slow. The meat is too raw; the meat is too overdone.

These guests are difficult to please. They thrive on criticizing and enjoy the attention they receive as a result, not the resolution of the problem, which removes them from the limelight.

Resolve their complaint quickly and as privately as possible. Be pleasant, professional, courteous, and efficient. The faster the situation can be resolved, the better for all. Whenever possible, promise to check back with the troublemaker to make sure all is well. He or she is looking for attention, so do whatever you can to make the guest feel special.

Top-of-the-Line Seeker

This is the guest who seeks authority, who will only speak with the highest-level manager at the hotel or in the restaurant, the guest who wants the person "at the top."

Many hotels and restaurants instruct their department managers to assume this level when speaking with authority seekers. There is the case of the disgruntled guests at the front desk of a resort hotel, dissatisfied with the room assigned to them, who asked to speak to the general manager. The front desk manager was informed of the situation by the front desk clerk. The front desk manager walked out from behind the front desk, removed the couple to a seating area away from the front desk area (and away from the majority of the remaining customers in the lobby area), and told them, "I am the manager. How can I help you."

Assume the authority, empower yourself, and take personal responsibility for the problem. Ensure the guests that you do have the authority to solve their problem; then proceed to solve it to their satisfaction.

If you do need to go to your manager or the general manager to solve the guests' complaint to their satisfaction, tell them that you will work out the details of a satisfactory solution with your assistant and you'll get back with them tomorrow. Do what you can to ensure a pleasant stay that evening and discuss the incident with your "assistant," the general manager, in the morning. Using this technique keeps you as the guests' "complaint solver." You have taken personal responsibility to solve their problem and are their personal contact. You can resolve the issues and ensure that they will be return customers.

Under the Influence

There will be times when a guest is under the influence of alcohol or other drugs that influence motor skills, including the ability to react and to walk surely. Guests under the influence of alcohol may be staggering, might miss the edge of their chair when trying to sit down, might miss the table when trying to replace a glass. Such drugs also affect the ability to think clearly, and these guests may be unruly, offensive, loud, belligerent, or even silly. They usually will argue with anyone who tries to cut off their liquor supply if they are in a bar or restaurant, stating matter-of-factly that they are not drunk. If a guest is under the influence of other drugs, or a combination of alcohol and drugs, their reactions might range from being quiet and withdrawn to being unpredictable and unruly.

The first course of action is to remove the guest who is under the influence of alcohol/drugs to a quiet and relatively secluded area, perhaps a back section of the lounge or a far corner of the lobby. The second step is to ensure that the person leaves the premises in a manner that does not endanger the guest or others. Chapter Seven provides detailed guidelines for dealing with intoxicated guests.

The Challenger

The challenger is similar to the troublemaker in that they both want attention; the challenger, however, is not as negative or disturbing as the troublemaker. The challenger wants to challenge house rules and regulations, similar to the way a 2-year-old child continually asks, "Why?"

The challenger might enter the restaurant in a T-shirt and slacks demanding to know why he must wear a coat and tie. He's there to eat, isn't he? Why should it matter what he's wearing? He can pay the bill just the same in a T-shirt as he can in a coat and tie, can't he? And so on.

Another common, "But why?" example is that of requiring shirts and shoes at resort retail outlets. People come off the beach or lakefront and want to "dash in" wearing their swimsuits and nothing else for a quick soda, more suntan lotion, or another beach ball.

Rather than answer, "It's house policy," respond with a reason. Challengers will not be happy with "it's our policy"; they usually react in a more friendly manner when they hear you state why. Often, they already understand or know the house rules but are trying to "bend" the rules for their personal convenience. By taking the time to explain why the policy was enacted and is being enforced, you're giving personalized attention to the guest, which is a form of compensation for not being able to ignore the rules.

AT THE TABLE: BEYOND THE FIRST IMPRESSIONS

S i x

Food Service

There are many opportunities to form a guest's "first" impression. That impression creates an expectation for the rest of the experience at that location, whether it's a meal at a restaurant, a stay at a hotel, or a visit to a resort or park. Once the guests have made their reservation, arrived, been met and greeted, and have checked their coats and/or parcels, that first impression has registered. If it was a highly positive experience, it leads the guests to expect more of the same.

❖ ATTRACTING RETURN GUESTS

Many hotels, restaurants, resorts, and parks provide adequate or even good service. Good service, generally speaking, is what guests expect. Sometimes, a few things are wrong with the service—it might not be as timely or gracious as the guest might wish—there might, in fact, be several things "not quite right" but not quite "off" enough for the guest to complain to management. The guests will leave the standard tip, not say a word, and probably not return.

To be competitive in today's marketplace, guest establishments must be better than good. They must provide better-than-"good" service that goes beyond the basics; they must provide excellent service in every aspect to all guests. For example, when guests check into a hotel, the wait at the service desk should be minimal. Many hotels are trying various "advance check-in" services such as allowing business guests to check in on the "business floor" or "tower" guests to check in with a desk clerk at the "tower level." Hotels realize that guests don't want to stand in line more than 10 minutes; if the wait is longer than that, they'll seriously consider staying somewhere else during their next visit.

When assigning a room, the front desk clerk should be conscious of any special considerations the guest might enjoy. Does a woman who is traveling alone want to be on a low-level floor or close to the elevator? Does a physically challenged person desire a room with special amenities such as TDD equipment or an amplified phone handset? When handing a guest a room key, the desk clerk should never state the room number aloud. The goal should be to provide "better than basic" service in all these areas.

❖ REVIEWING THE BASICS OF RESTAURANT SERVICE

To provide a better understanding of some areas where restaurant service operations can offer superlative service, Chapter Six will concentrate on table service; Chapter Seven then focuses on wine and liquor service. Later chapters will cover unusual foods and beverages as well as policies that can create and maintain a unified, well-groomed, well-dressed staff.

After a party has been seated in a restaurant, someone on the service staff team should acknowledge the guests within one minute. All members of the service staff should be wearing smiles as well as being immaculately groomed, in impeccable uniforms. The restaurant should be spotless and litter-free. The restrooms, in particular, should be inspected hourly for cleanliness and restocked as necessary with soap, towels, and toilet paper.

❖ GOING BEYOND THE BASICS

Giving restaurant guests the best service they have ever had should be the goal of every employee, because such service has a positive ripple effect. If every employee provides better than basic service, the guests will leave large tips, making the service staff happy. Guests will return, making both management and the service staff happy, and guests will tell their friends about the establishment, making management, the service staff, and those friends very happy. This chain reaction marks the successful establishment.

Inspecting the Table

An area that guests immediately notice is the tabletop and chairs. Can they meet the following tests:

- Does the table appear symmetrical? Are the place settings balanced and, if an even number, opposite each other?
- Glass or ceramic centerpieces should be free of fingerprints and dirt and "splashed-on" food particles. Candles should be lit during evening dining hours. If the centerpiece contains flowers, are they fresh and in good condition?
- Inspect the table linen. Is it clean, pressed, without holes?
- Do the chairs match? Are they clean? Check upholstery for crumbs or spills.

Checking the Flatware

Look at the flatware: Is each piece spotless, without water marks, fingerprints, dirt, or caked-on food? Flatware should be wiped dry after leaving the warewashing station to ensure that no water marks remain. If the service staff handles flatware by the waist (the middle portion of the piece), there will be no fingerprints either at the top or the bottom. Further, all flatware should be gathered and handled inside a cloth towel or napkin; staff members should avoid touching it with bare hands.

These steps will ensure that the flatware is polished, clean, and free of fingerprints. As the service staff dry and polish each piece (immediately after warewashing while it is still damp), they can check each piece for cleanliness. Any pieces with caked-on food (eggs are particularly hard to remove) can be immediately returned to a soaking bin for further warewashing.

Silverplated flatware may become tarnished. A quick method for the service staff to quickly brighten tarnished silverplate is to mix a solution of hot water, salt, lemon juice, and a ball of aluminum foil. Put all ingredients into a container and immerse the tarnished flatware for about one minute; remove, rinse, and polish. This is a quick "fix" to an ongoing problem that any service staff member can easily learn.

By having each staff member check and polish the silverware at his or her station prior to service, the whole restaurant will have polished silverware for that meal period. Repeating this process before every meal period provides a simple method for ensuring polished silverware. If there is a burnishing machine on the premises, so much the better. It will handle the routine maintenance and polishing of the silverware. On a day-to-day basis, however, there needs to be a system that guarantees the guests will have polished silverware (other than breaking out newly purchased pieces every day). Having polished silverware at every place setting, every meal period, is one of the difference makers that will separate a restaurant from others that are just "okay."

If a guest knocks a utensil off the tabletop, a service team member should replace it discreetly but immediately. The replacement piece is brought to the table resting on a cloth napkin on a plate, and the service team member removes it from the plate touching only the waist. (Knives are placed with the cutting edge pointing inward, toward the center of the place setting.)

Procedures for Checking Glasses

Look at the water glasses. Are there water marks? Are there fingerprints? Are there any chips or cracks?

For guests seeking beyond-the-basics service, any of the preceding flaws are signals to try another restaurant the next time they dine out. Glassware, like flatware, should be free of spots and fingerprints. If there are spots, the rinse water may not be as clean as it should be; or chemicals in the rinse water could be streaking the glassware. Or, an improper mix of washing and sanitizing chemicals might lack the sheeting action that makes the water "sheet" off the glass without streaks or water marks.

As a last resort in the dining room, the service staff can steam then wipe (using a clean towel or napkin) any spots and/or fingerprints off glassware. Only water will remove water marks, therefore a dry towel or napkin will not work; it must be damp or the glass must have condensed steam on the inside of the bowl. Wiping the bowl and stem dry should remove the fingerprints.

While handling glassware, the service staff members should never touch the rim of the glass, where the guest's mouth will touch. Fingerprints on the top portion of the glass denote a sanitary concern. (Who knows where those fingertips were 2 minutes ago, 10 minutes ago? Were they brushing through the server's hair? Picking up a fork that had dropped to the floor?)

Always handle stemware glasses by the stem. If the glass doesn't have a stem, it should be handled by the bottom 2 inches of the glass. This method isn't only for clean glasses and not only for handling glassware when guests are present. Servers must follow these same methods at all times—in the kitchen, when setting the tables, when clearing the tables, even when out for dinner as guests themselves, or while relaxing at home with friends. This behavior will make the correct handling of glassware a truly automatic habit. Only then can servers feel confident that they won't put their fingers into the tops of glasses when they are busy or under stress. Guests who observe servers handling glassware properly will feel secure regarding the glasses that they are using.

Handling stemware by the stem and other glassware by the bottom two inches helps to ensure that guests are drinking out of completely sanitized glasses. Warewashing isn't enough; the service staff's handling influences how clean each piece really is when it reaches the guest.

Serving Beverages

If one person in a party asks for water, either water should be brought for everyone in that party or the server should ask whether anyone else desires water before getting it for the requesting person.

In either case, it should be brought immediately. Once water has been requested, it should not have to be asked for again. Water should automatically be refilled throughout the meal until the dessert course. After the dessert has been served, the water glass is allowed to go empty. If guests must ask for water refills, that too, is a sign of average service, of the service staff not going beyond the basics.

Beverage service should be from the right side of a guest, the server using his or her right hand. This permits the server to face the guest while serving, so the guest can speak to the server if necessary. In addition, it's polite body language; the server's back would be to the guest when using the left hand to place a drink at the guest's right side.

Arguably, the person could use the left hand and serve the beverage to the guest's left side, in which case the server would be facing the guest. However, this ignores the convention of placing beverages at the tip of the knife on the guest's right side. This convention evolved through years of custom since most people are right-handed (approximately 70% of the population). Placing drinks at the right side of the place setting, approximately at the tip of the knife, minimizes the hand movements of most guests. If there is no knife, beverages are placed on the right side of the guest approximately 12 inches from the edge of the table, where a guest's hand would be if a guest placed his/her elbow at the table edge and reached forward.

Should a person serve from the left using the left hand, placing the beverage on the guest's right side, it is necessary to reach across the front of the guest's place setting, putting the server's body in the path of table conversation and in the guest's face. Therefore, unobtrusive, sophisticated, upscale beverage service requires serving beverages to the right side of a guest using the right hand. The beverage tray, necessarily, is held in the left hand. It is absolutely irrelevant whether the server is "right-handed" or "left-handed."

The server of beverages should walk clockwise around the table if possible. Servers walking in a counterclockwise direction (often observed in establishments not yet "beyond the basics") are necessarily walking backward if they are serving to the right side of the guest with the right hand. While walking backward may be one way to get around a table, it is certainly not the most efficient, fastest, nor the safest route.

To summarize, beverages should be served from the right using the right hand, traveling in a clockwise direction around the table, if possible (not at a booth, wall table, etc.). Servers should handle stemware by the stem and tumblers by the bottom 2 inches of the glass.

How to Take a Food Order

It's time to order. Does the server begin by taking the orders of the women in the party? Does the server move around the table, if possible to take each person's order? Or, does the server stand at one spot as though glued to the floor, requiring guests to shout their order above the noise of background music, the talking of other patrons, and so on? The server's choices help determine whether service is average or beyond the basics.

Serving the Meal

The food is arriving. Is each plate placed from the left side of the guest by a server using the left hand? Is each plate placed without anyone asking, "Who gets the . . . ?" Is each plate placed without the server's arms passing in front of or reaching over guests?

A server does not need to inquire "Who gets . . . ?" in an upscale restaurant—there are many methods to assure knowing who gets what. (A common seat designation system is detailed in our book, *Food and Beverage Service*, (New York: John Wiley, 1990) page 91.)

A server must never reach over a guest to serve another guest if there is another way to serve that guest without doing so. Generally, such servers are just being lazy because they don't want to walk around the table to serve their guests; they stand in one spot and reach, handing the plates to the various recipients. This "lazy method" brings to mind the familiar question, "Are you tipping for service or just for food arriving at your table?"

To provide excellent, customer-based service, the server should move so unobtrusively that the guests at the table barely notice. The server first serves the women's plates and then the men's, walking around the table serving each guest from the left side, holding the plate of food in the left hand, traveling in a counterclockwise direction. (See *Food and Beverage Service*, p. 116, for details of this technique.)

If guests are sitting at a booth, against a wall, or with a pillar blocking access, a server might be forced to reach over someone while serving or clearing. These are necessary and understandable exceptions. The server should always say, "Excuse me," while reaching over a guest to serve another guest's plate.

Some restaurants have French plate service without incorporating complete French service (using the réchaud, the gueridon, etc.). In a restaurant with French service, the server will use the right hand to serve and clear plates and will serve the plates from the right side of a guest traveling in a clockwise direction around the table whenever possible. Either French service (serving food from the right side of a guest using one's right hand) or American service (serving food from the left side of a guest using one's left hand) is correct as long as it is unobtrusive, skillful, and consistent. An upscale restaurant can use either style with confidence.

Providing Condiments and Refills

Have you ever received a hamburger or french fries but no ketchup? Received your coffee or tea but no cream or sugar? Received the lobster tail but no drawn butter?

Servers should ensure the guests at a given table have all condiments that may be needed prior to the food or beverage arriving. A guest's food or beverage should not be cooling while the server scrambles to find the ketchup, the cream or sugar, or the drawn butter. While it might take just a moment to procure the condiment and bring it to the table, if the server forgets to do it immediately and instead goes into the kitchen to retrieve the next table's order, it might be 10 or 15 minutes before the server remembers (or is reminded by the guests) and finally retrieves and delivers the needed condiment(s). Meanwhile the food is cold, the coffee or tea is cold. And the tip for the server?

What about coffee refills? Are they automatic? At least one coffee refill per guest should be automatic. Some restaurants offer unlimited coffee and should refill your cup every time it is near empty. Other restaurants have instituted the use of thermal containers and place a pot on each table so guests can refill their own cups. This is an effective and popular method at family-style restaurants.

Clearing the Plates

It's time to clear the plates. Some restaurants dictate that no plates be cleared until all diners finish their meals, other restaurants encourage their servers to clear as diners finish as long as at least two diners are still eating and have their plates in front of them.

Both methods have their place. Usually, upscale restaurants that are trying for two turns or more per meal period will clear as guests finish as long as at least two guests are still dining. At truly upscale restaurants with only one seating per meal period, all plates are left until everyone in the party has finished; then all plates are cleared. Often two servers or bussers will clear a table simultaneously to provide expedient service.

Clearing the plates should be unobtrusive. Ask the following questions:

♦ Did the server or busser reach over guests to clear the plates?
♦ Did the server or busser reach out with one arm and wait for the guests to pick up the dinner plates and hand them across the table?
♦ Did the server or busser stack plates at the table, in front of guests?
♦ Did the server or busser stack bread-and-butter plates (B&B plates) atop dinner plates in front of guests?
♦ Did the server or busser put clean, unused silverware onto plates before removing them?

If any of the preceding occurred, it wasn't beyond-the-basics service. It was terrible service.

Clearing of plates should be from the right by the right hand, traveling in a clockwise direction around the table. This method is used whether the serving of the plates was from the left side of the guest (American service) or from the right side of the guest (French service).

Plates should be cleared without stacking anything in front of guests and without stacking anything onto anything while it's on the table and guests are there.

The box "Procedure for Clearing Tables" provides guidelines for handling most situations.

PROCEDURE FOR CLEARING TABLES

1. Pick up the dinner plate from guest A (a woman, if a woman is present at the table). Pick it up from her right side using your right hand. (If a woman is not present, begin with a man.)
2. Place the dinner plate in your left hand, holding your left hand in the aisle space.
3. If there is any unused flatware on the right side of guest A's place setting, reach back to the table, on the right side of the guest and quietly remove each piece of flatware. Bring each piece out to the aisle space and place atop the dinner plate in your left hand.
4. Move to guest A's left side. Quietly reach, using your right hand, to the B&B plate. Remove the B&B plate and B&B knife, if it's atop the B&B plate, in one motion. Remove the B&B plate and knife to the aisle space and place atop the dinner plate.
5. If there is any unused flatware on the left side of guest A's place setting, reach back to the table and quietly remove each piece of flatware. Bring each piece out to the aisle space and place atop the dinner plate in your left hand.
6. Move clockwise to the next guest and begin with his or her dinner plate. Stack the dinner plate atop the dinner plate from guest A if possible, or place all of guest A's plates and flatware on a service tray prior to beginning guest B.
7. Repeat steps 3 through 7 until all guests are cleared.

Each guest's place must be completely cleared before moving on to the next guest. The server shouldn't remove all the dinner plates, then come back for all the B&B plates, then come back again for all the unused flatware. That would necessitate three intrusions on each guest to clear the table.

Even three separate clearings, however, would be preferable to having a server stack plates and flatware in front of the guests at the table. The best option is to have one intrusion per guest, allowing for fast, expedient "beyond the basics" service while clearing the table.

Wine and Liquor Service

Cocktail service is another difference maker. Have you ever entered a lobby lounge or cocktail lounge only to be seated and have no one approach you for an order? Or, have you given your order to a cocktail server only to have the person disappear for about a half hour?

Cocktail service should be quick. The servers must be friendly because they are often serving guests who would rather be in the restaurant, dining. If the cocktail lounge is serving as a "holding tank" for the dining room, some patrons may become impatient and irritable. To prevent this, a cocktail server should acknowledge each party as soon as the guests are seated, or within one minute thereafter.

The cocktail order should be taken soon thereafter, and complimentary snacks, if served, should be placed on the table. It is imperative to get the first round of beverages to the guests quickly. Once the first round of beverages has been served, the pace of service can slow somewhat, allowing the server to make the round of his or her station and provide refills to other tables.

With the advent of "healthy lifestyles," many people are choosing nonalcoholic beverages. This chapter reviews traditional cocktails, beer, and wine. Chapter Eight describes nonalcoholic beverages, especially fruit and vegetable drinks and bottled waters.

❖ TRADITIONAL COCKTAIL SERVICE

Occasionally a guest will order a martini, and the server will leave only to return a moment later to ask if that will be a gin martini or a vodka martini. Next, the server comes back to ask if that will be "up" or "on the rocks." And yet a third time, the server returns to ask if the guest wants an olive or a twist for the garnish.

Hopefully, a review of this chapter will enable future cocktail servers not to make such errors, costly in time even if the guests are good-natured about the learning process.

When taking the cocktail order, the server may place cocktail napkins on the table unless the lounge is outdoors and the wind might blow them away. This is often used as a signal to other servers that the table's order has been taken.

When cocktails arrive, they should be placed on a beverage napkin unless there are linens on the table, in which case the napkin becomes optional. The server uses the right hand to place the drink by the guest's right side. If possible, the server should travel in a clockwise direction around the table.

Cocktails are usually made by the bartender and garnished by the server. In some operations, the bartender creates the drinks and garnishes them; in other operations, the server presents the glass to the bartender when ordering the cocktail and garnishes it after receiving the liquor for the cocktail. The glassware and the cocktail garnishes of olives, cherries, and cut fruit would then be on the server's side of the bar. Often, ice and mixers—ginger ale, soda water, and tonic—are also on the server's side. The bartender may only dispense the liquor and prepare the drinks that require mixing. For this type of operation, the server must know:

♦ The proper glass for each drink, according to the standards of the operation.
♦ Appropriate garnishes for each drink.
♦ The appearance of the drinks ordered, so only the drinks ordered are taken and customers are served their exact order.

These items are covered in detail later in this chapter. It is important to establish standards for preparing drinks so there is consistency in the beverage service. When guests order a particular drink, they expect it always to look and taste the same.

Furthermore, following established policies, procedures, and measures helps with forecasting glassware, garnish, liquor, and mixer usage and purchases. Setting price levels, especially when they are combined with electronic point-of-sale (POS) systems, can substantially increase servers' efficiency, enabling them to serve more drinks in the same shift. This translates to more tips for the servers, increased sales and profits for the operation, and better satisfied customers.

❖ COCKTAIL SERVICE BASICS

Cocktail service involves many detailed operations and specialized knowledge. The following sections provide information about service procedures, the kinds of drinks served most often, and methods of payment.

Initial Table Service

Cocktail service, whether in a dining room where guests will be having lunch or dinner or in a cocktail lounge, must be prompt and courteous. When guests are seated, the first thing they usually ask for is something to drink—water, an alcoholic beverage, a soda, and so on.

If several tables of guests arrive at once, the cocktail server should first meet and greet the guests, saying, "Good afternoon [evening], I'll be right with you . . ." This acknowledges the guests' presence and lets them know who will be serving them.

Ask the guests whether they will be dining or having only drinks so you will know whether they will (probably) have only one round and then move to the dining room or may be in the lounge for a while, having several rounds.

When many tables of guests arrive at approximately the same time, take orders for several tables at once; then go to the computer terminal (if using one) and enter all the orders. After entering the orders, proceed to the bar to obtain the drinks—as many as can be carried. If electronic terminals are not used, place orders directly at the bar.

Always serve an entire table at once. If possible, serve two or three tables after every trip to the bar. When serving drinks, remember these rules:

1. Beverage trays are always carried in the left hand.
2. Drinks are always served with the right hand, from the right side of a customer if at all possible. If you are serving a banquette table, serve from the outside (aisle) in, toward the banquettes.
3. Glassware is always handled by the stem if there is one, or by the bottom third of the glass if there is no stem.

If there are tables waiting for drinks and tables waiting for food, stop by the tables waiting for food and explain to them that you'll bring their food in just a few moments. This reassures the customers, who now are assured that their server is aware of them and realizes they are waiting for food. Serve drinks first, then food.

If there is a cashier (server banks are not used) and several parties are ready to pay their bill or "settle their check," collect the checks and money or credit cards from all the "ready-to-pay" tables at the same time. This will reduce the length of time you must wait in line at the cashier.

Organization is the key to satisfied customers, high sales, and good tips. Organize your station: Collect and serve drink orders for several tables in one trip; order and pick up the food for several tables at one time; and close out several checks at the cashier at one time. This optimizes time and motion to your best advantage.

Ordering Cocktails from the Bartender

At many establishments, cocktails, or mixed drinks, are mixed by the bartender. The cocktail server, however, may be responsible for choosing the glasses, garnishing the drink, and sometimes even putting in the ice.

When arranging glasses on a beverage tray for bartenders to fill, servers at some establishments group all like liquors together (all vodka drinks together, then all rum drinks, etc.).

Other establishments allow the servers to place the glasses on the tray in any order they wish and call them in any order. If this is

---

SERVER PROCEDURE FOR PREPARING COCKTAILS

1. Place the glasses required for the drinks on a beverage tray.
2. Ice them; garnish them. Arrange them on your tray so they can be called to the bartender.
3. Present the bar duplicate check (dupe) (obtained from the electronic cash register) or bar slip to the bartender if there is not a remote printer at the service bar station. This allows for checks and balances within the operation—a control over what was poured against what was ordered and paid for by customers.
4. Call the drinks to the bartender. Some operations establish a calling policy; for example, the server may call the drinks looking at the tray from left to right. The bartender will then fill the glasses with liquor from his or her right to left.
5. Obtain sodas, water, stirrers, straws, and final garnishes.

the case, for optimum speedy service, arrange the drinks so that when leaving the service bar to serve the drinks, the women's drinks for table 1 will be in front and thus the first to be delivered, the men's drinks for that table will be placed behind those, then the women's drinks for table 2 become most accessible on the tray, then the men's drinks for table 2, and so on.

Commonly Used Terms

Certain phrases are used over and over again when ordering cocktails. To make order taking faster and easier for servers, common abbreviations have evolved. These abbreviations make it simple for one server to take the order and hand it to another to pick up if a station becomes swamped or servers are working in teams. For these reasons, use the same abbreviations within a restaurant. The following abbreviations and terms are standard in most bars and cocktail lounges.

WW	White wine
RW	Red wine
X	With ice (on the rocks)
↑	Without ice (straight up)
TW	With lemon twist
OL	With olive
PUS	Martini *P*itcher + *U*p glass + clear plastic *S*tirrer

Straight Up. A mixed drink (e.g., a martini) served without ice, in a stemmed glass.

On the Rocks. A drink, either a mixed drink or a liquor, served in a squat glass, poured over ice. Iced drinks must be served immediately so they do not become diluted.

Martini. A cocktail consisting of liquor and vermouth that requires two server questions: "Up or on the rocks?" and, "Olive or twist?" Some operations also suggest asking: "Gin or vermouth?" Gin is the default. A martini ordered with an onion garnish is called a "Gibson."

Tall Drink. A drink made by combining a liquor and a mixer (e.g., gin and tonic), which is served in a tall glass, usually garnished with lemon or lime.

Stirrer. A disposable utensil, usually resembling a plastic straw, placed in all drinks "on the rocks," served in "rocks" glasses or highball glasses.

Highball. A drink made with liquor (usually whiskey) and water or a carbonated beverage (e.g., scotch and soda). It is served on the rocks with a stirrer in a highball (tall) glass.

Sour. A cocktail consisting of liquor (especially whiskey), lemon, lime, sugar, and sometimes water or soda. It is shaken with cracked ice and served in a "sour" glass. It is garnished with a cherry and orange slice. Any drink resembling a sour (e.g., a daiquiri) also takes a cherry.

Old Fashioned. A cocktail consisting of whiskey, sugar, and bitters served over ice cubes in a "rocks" glass.

Split. With reference to mixers, a small bottle holding about six ounces of mixer. It ensures quality and portion control to the guest, who receives the whole bottle (without the cap) at the table and dilutes his or her drink to the degree desired.

Spilled Drinks. A reference to the tendency of tall glassware to be unstable. Slight spills at the base are wiped without lifting the glass.

To steady topheavy champagne glasses, pilsner glasses, and oversized wineglasses on a beverage tray, servers often spread the fingers of the right hand atop the bases of several glasses.

Placement. In this context, the proper arrangement of glassware on the table, which is directly above the entrée knife. If there is a water glass on the table, the cocktail should be placed just below and to the right of the water glass.

Service Bar Payment Procedures

Some bars use a cash bar system, whereby servers pay for the drinks as they are delivered to them by the bartender. In these systems, the servers have "server banks." The server takes the guests' order, goes to the bar, and calls the order to the bartender, who makes the drinks. The server pays for the drinks and delivers them to the guests. The server can either collect for the drinks after every round (COD), or can "run a tab," in which case, the party will "settle up" or pay when ready to leave.

Should a group walk out without paying the tab, the server usually must absorb the loss for the cost of those beverages. Therefore, if the server does not recognize the guests (they are not "regulars" or hotel guests, whose charge can be added to their room charge), it is wise to collect for each round of drinks as the drinks are delivered to the table.

With a cashier, point-of-sale, or computerized system, cocktail servers usually must present either a written or computer-generated beverage "dupe" (duplicate check) to bartenders in order to receive wine, spirits, beer, cocktails, or other alcoholic beverages.

Pricing and Payment Policies

The server usually charges for drinks by designating, on a guest check in some manner, the names and prices for each drink ordered. Some operations have manual systems whereby each drink is assigned a price and guests are charged accordingly.

Some establishments, using a computerized system, have a "price/drink button" or "price-look-up" (PLU) number for every drink. In computerized establishments, servers will, after a few weeks, memorize the most frequently ordered drinks and their respective codes and know which keys to push for which drinks. Common price-look-ups thus become memorized quickly and the remainder can be found on a list. Some of the look-up sheets are arranged alphabetically, others by kind of drink (e.g., rum drinks, scotch drinks).

Some establishments group drinks by type. The servers memorize the drinks in each category and then charge for that category or level. For instance, if Screwdrivers and Bloody Marys are categorized as Level 1, and one of each has been ordered, the beverage dupe would charge for two Level 1 drinks. Even easier are "touch-screen" POS systems, whereby servers don't have to memorize or "look-up" anything.

In categorizing drinks within price ranges (for the convenience of the service staff, and ease of accountability), each level represents the price for several drinks; rather than memorizing a price for every drink, the servers need only remember that drink's level, or category. Should drink prices rise, the servers need only remember the new prices for each level (see Sample Price Levels chart).

SAMPLE PRICE LEVELS		
Price	*Price Level*	*Type of Drink*
$3.50	1	Bar liquor
4.00	2	Call liquor
4.50	3	Call plus
4.75	4	Premium
5.25	5	Aperitifs, Cognacs, Specialty drinks

Computerized operations are even easier—the servers simply press the appropriate key and the computer recognizes the correct price. The computer can also make it easier to track inventory and control beverage usage.

The accounting department determines which drinks to include in each price level by calculating the cost of liquor, mixers, and garnishes as well as a breakage factor for the glasses (specialty drink glassware is often expensive). Management's markup and profit are then added in to establish the price. Determining levels and "house recipes" for each drink is a time-consuming job; once done, however, it can save a lot of time later when tracking the profitability of the bar.

When an establishment uses a level system, it is imperative that the bartenders adhere to the house formulas, recipes, and "pours." If drink levels are computed on 1-ounce pours, bartenders who dispense $1\frac{1}{4}$-ounce pours are invalidating the pricing structure.

MAJOR DRINK CATEGORIES

Levels

1   Bar Liquor
2   Call Liquor
3   2 Calls or Premium
4   2 Premiums or Special
5   Special Drinks

Word Keys

Soda	Juice
Perrier/ Bottled water	Glass wine
Glass—Wine 1	Glass—Wine 2
Glass—Champagne	Virgin drink
Imported beer	Domestic beer
Special beer	Full carafe
Half carafe	

Miscellaneous (MISC) Keys

Soda	Beer
Liquor	Food
Wine	

Pricing Beverages at a Point-of-Sale Terminal

Prices for beverages vary, and so do the ways of entering those prices on a computer terminal.

COMPUTER TERMINAL KEYING CONFIGURATIONS

Word Keys:

Glass Wine	Imported Beer	Domestic Beer	Soda

Liquor Price Keys:

Liquor $3.50	Liquor $4.00	Liquor $4.50	Liquor $4.75

Level Keys:

Liquor Level 1	Liquor Level 2	Liquor Level 3	Liquor Level 4

Actual Price Keys:

Misc Wine	Misc Liquor	Misc Beer	Misc Soda

There are any number of ways to enter the data. On computerized point-of-sale terminals, the keyboard may have designated keys for each level (a "Level 1" key, a "Level 2" key, etc.), price keys, or keys/entries for each drink, by name.

Common Level 1, 2, and 3 drinks, and their prices, are detailed in the Price Level Chart for Cocktails. Since each establishment will have its own pricing structure, levels are listed rather than actual prices. An establishment would then list drinks by level and have a separate chart listing the prices for each level:

Level 1 drink	$3.50	Level 4 drink	$4.75
Level 2 drink	$4.00	Level 5 drink	$5.25
Level 3 drink	$4.50		

A drink with a different price might be listed under MISC. "MISC" is most often used for unusual combination drinks, such as a Long Island Iced Tea, or an unusual variation of a typical drink (e.g., a White Russian with a half-shot of Frangelico). If prices rise, the management can post the new prices for each level and not have to change the detailed chart, listing all drinks. Since that chart is based on the cost of each drink but groups together those drinks using similarly priced liquors, the drinks in each level should remain fairly constant even with increases of liquor prices.

After establishing a system for pricing all bar drinks, the establishment should post the required procedures and create a handbook or handout for the service staff.

A useful addition to such a handout is a list or chart designating the proper glassware to use for each kind of drink. Guidelines for this type of information are included later in this chapter.

When analyzing beverage operations and establishing prices, controls, and checks and balances, managers should keep the following factors in mind:

♦ *Ease of Use.* The system must be easy for the servers to learn and enable them to take drink orders quickly and accurately. There must be only one way to order each drink, and only one price for that drink.
♦ *Control of Inventory.* Par stocks must be established and a system implemented for control of inventory—what has been released from storage, what has been used.
♦ *Checks and Balances.* Physical inventory used must match paper usage, as tracked by servers' duplicate drink slips, whether handwritten or computer driven. Variances must be tracked, and excessive percentages (as determined by the operation) require management intervention. Ability to track each order to the server is necessary, whether by name or identification number.

Control points to check include use of the "misc liquor" key. If this category is available, management must question the bartenders as to what was poured for those drinks. Another control point is "errors." If a server uses an improper category for a drink, and verbally tells the bartender what the drink "should have been," the bartenders cannot pour the drink until the computer record is corrected.

When establishing a beverage system, categories such as those suggested in the Major Drink Categories chart can be established, in addition to "levels."

The following information is organized for use on electronic keyboards/point-of-sale systems since they are so prevalent. They provide ease of use for the servers and bartenders, as well as an accurate "paper trail" of beverages ordered and sales to compare against physical inventory.

Electronic point-of-sale terminal policies and procedures are available from manufacturers of such equipment. Many companies offer such equipment, and several vendors should be evaluated prior to purchase and installation to ensure that the system you choose is designed to meet your needs. Some of the easiest to use systems use touch screens rather than keyboard entry and regular paper rolls of paper rather than formal prenumbered checks. There are hand-held data-entry devices that can communicate directly to the kitchen or the service bar. Check with the vendors regarding these and other options.

❖ # A PRIMER OF ALCOHOLIC BEVERAGES

House Wines and Champagne

Wines will be discussed in detail later in this chapter. At this point, however, it is useful to keep in mind that guests often request house wines or champagne.

Wines

When customers request a glass of wine they often desire a light, drinkable wine and are satisfied with a "house wine," usually a glass of white or red jug wine. American jug wines are designed for popular taste—not too heavy, not too much bouquet or astringency, not too sweet, not too dry. Jug wines are made to be consumed as soon as bottled. They can be from any country—restaurants often use a variety of American, Italian, and French jug wines.

A house white wine is usually a light, dry wine with no special character or bouquet, acceptable to a wide variety of people. A house red wine is usually a light- to medium-bodied wine with a moderate bouquet and no distinguishing characteristics, also acceptable to a wide variety of guests.

All service staff should know the brand names of the house wines being poured. Restaurants with a growing number of wine connoisseurs as patrons often serve a "wine of the day" or have several wines available "by the glass." If the wines of the day or wines by the glass are changed on a fairly regular basis, the names of the current house wines being poured should be posted by the service bar where the servers and even patrons can easily refer to the list.

Champagne

A house champagne is usually a sparkling wine. Spanish sparkling wines are frequently chosen because many are fermented using *méthode champenoise* (champagne process), the procedure followed in making true champagne. Méthode champenoise indicates that the second fermentation was in the same bottle as that sparkling wine is shipped in and served from. For méthode champenoise sparkling wines, the second fermentation period is at least 18 months, the same time allowed in Champagne, the district in France where true champagne is made.

Being aged for 18 months before being released for consumption gives the wine its sparkle, a nice flavor, and small, well-released bubbles. Generally a *brut* (dry) or *extra dry* (slightly sweeter than brut) sparkling wine is used as the house champagne since these are agreeable to most people and work well for champagne drinks such as mimosas and kir royales.

House Specialty Drinks

When choosing categories of drinks, management must decide whether to serve specialty drinks such as ice cream drinks and frozen daiquiris. Frozen drinks require providing a small ice cream freezer and blenders in the bar area. If frozen drinks are not served, the staff must be aware of this policy so they don't accept guests' orders for such drinks only to have to return and ask for a reorder.

If there is a house substitute for certain drinks or types of drinks, this information must be disseminated to the staff. For instance, if an operation doesn't choose to serve a variety of frozen or blender drinks, but wants to offer a reasonable alternative to customers wishing such a drink, the operation might prepare large batches of one blender drink in advance and offer that as the "house specialty." Rather than serving frozen daiquiries in all flavors, taking much of the bartenders' time for individual drinks during peak service periods, the bar might offer the "house specialty strawberry daiquiri," prepared in advance from fresh strawberries and a bit of sour mix with rum and cracked ice. Large batches of this specialty drink would be prepared by the kitchen staff or a bar attendant during the slow periods of the day for use during the evening, brunch, or other busy periods.

Specialty drink information should be relayed to the staff frequently as the house specialties change. In addition, suggested selling techniques should be reviewed at staff meetings so the staff can market the "specialty strawberry daiquiri" or other house drink, and the staff members can be given nonalcoholic samples to try.

If a house drink is created, make sure it is exceptional because the reputation of the restaurant will be associated with it. If fresh strawberries are advertised, use them! If the specialty house drink is truly exceptional it will alleviate wait times at the bar during peak times and facilitate customer service.

Standard Alcoholic Drinks

The chart on pages 106–115 is an overview of standard alcoholic drinks, categorized by the predominant liquor(s) used in the drink. Each listing states:

♦ The name of the drink.
♦ A typical pricing structure (based on the pricing levels described earlier, level one being the least expensive drink category).
♦ The predominant ingredients used.
♦ Questions the server should ask the customer.
♦ Notes on the drink.
♦ The typical glassware used for serving the drink.

The Basic Liquors chart (p. 116) shows commonly requested brands of liquors for use in these cocktails. These two charts provide a basic guide to the most popular alcoholic drinks ordered in food service operations.

Some Other Bar Services

Sodas

For soda service, use a highball or red wineglass and serve the soda with a straw. There are two popular types of service: club service and fountain service (see box on p. 117). Club service entails the use of individual soda bottles or cans for each person. Each guest receives the entire bottle or can and may receive two or more pours into a glass of ice. Fountain service entails the bringing of a filled soda in a glass to the guest. Each refill in fountain service is charged for separately. Several popular brands of sodas are shown in the following list, including some of the popular clear sodas.

7–Up	Diet Pepsi
Ginger Ale	Fanta
Coke	Crystal Pepsi
Tab—Serve with lemon wedge.	Nordic Mist
Club Soda	Snapple
Tonic Water—Serve with lime wedge.	A&W Root Beer
Sprite	Squirt
Pepsi	

GIN, RUM, WHISKEY

Martini

Level   1

MAJOR INGREDIENTS   Gin plus dry vermouth

ASK   "Up or on the rocks?"

"Olive or twist?" (sometimes w/lime)

NOTES

Ask, "Gin or vodka?" (Gin is the default)

Ask Europeans about preferred liquor, as they sometimes mean straight vermouth when they ask for a martini.

Gibson   Same as martini, but garnish with *cocktail onion*

GLASSWARE   On the rocks: Rocks

Up: Martini pitcher, clear plastic stirrer, up glass

GARNISH   Olive or lemon twist (twist)

Manhattan (Sweet)

Level   2

MAJOR INGREDIENTS   Rye plus sweet vermouth

Perfect Manhattan   Half sweet vermouth; half dry; garnish with twist

Dry Manhattan   Rye and dry vermouth; garnish with twist

ASK   "Up or on the rocks?"

GLASSWARE   Same as for martini

GARNISH   Cherry

Rob Roy (Sweet)

Level   2

MAJOR INGREDIENTS   Scotch and sweet vermouth

Perfect   Same as for Manhattan

Dry   Rob Roy

ASK   "Up or on the rocks?"

GLASSWARE   Same as for martini

GARNISH   Same as for martini

7&7

Level   2

MAJOR INGREDIENTS   Seagram's 7 and 7-Up

GLASSWARE   Highball

GARNISH   None

Rum & Coke

Level   1

MAJOR INGREDIENTS   Rum and Coca-Cola

GLASSWARE   Highball

GARNISH   None; lime if guest asked for a Cuba Libre

Old Fashioned

Level 2
MAJOR INGREDIENTS   Rye or bourbon whiskey, bitters, sugar, orange slice, cherry, splash of club soda
GLASSWARE   Rocks (Usually served on the rocks)
PROCEDURE   If servers muddle: get glass.
Put in bitters, sugar, orange, cherry.
Muddle with a small stick or back of a spoon.
Add ice. Give to bartender.
Bartender pours liquor.
Top with club soda; add stirrer.
GARNISH   Bitters, sugar, orange, cherry

Presbyterian

Level 1
MAJOR INGREDIENTS   Rye, bourbon, or other "call" liquor; soda or ginger ale
GLASSWARE   Highball
GARNISH   Twist

Americana

Level 2
MAJOR INGREDIENTS   Campari, sweet vermouth
GLASSWARE   Rocks
GARNISH   Twist

Negroni

Level 2
MAJOR INGREDIENTS   Campari, sweet vermouth, gin
GLASSWARE   Rocks
GARNISH   Twist

Sour

Level 1
MAJOR INGREDIENTS   Rye, Scotch, or vodka plus sour mix
NOTE   If other liquor desired, then Level 2 Liquor: e.g., Amaretto Sour, Apricot Sour, Midori Sour (Melon-flavored liquor)
GLASSWARE   White wineglass if on the rocks
Pitcher, up glass, stirrer if straight up
GARNISH   Orange slice, cherry

Bacardi Cocktail

Level 2
MAJOR INGREDIENTS   Bacardi rum, sour mix, grenadine (a red, pomegranate-flavored syrup)
NOTE   All else same as for sours.

Daiquiri

Level   3
MAJOR INGREDIENTS   Rum, sour mix, sugar
GLASSWARE   Same as for sour
GARNISH   Lime
NOTE   Are frozen daiquiris available?

Fresh Strawberry Daiquiri

Level   4
MAJOR INGREDIENTS   Rum, fresh strawberries, sour mix, touch of sugar/
artificial sweetener
GLASSWARE   Champagne glass
GARNISH   Sugar rim, fresh strawberry, straw

Marguerita

Level   3
MAJOR INGREDIENTS   Tequila, Triple Sec, sour mix
ASK   "Salt on the rim?"
"Up or on the rocks?"
NOTE   Are frozen margueritas available?
GLASSWARE   Same as for sour
GARNISH   Salt, if desired; lime

Side Car

Level   1
MAJOR INGREDIENTS   Cognac, Triple Sec, sour mix
ASK   "Up or on the rocks?"
GLASSWARE   Same as for sour
GARNISH   Lime on rim of glass

Godfather

Level   3
MAJOR INGREDIENTS   Amaretto, Scotch
GLASSWARE   Rocks if on the rocks, or pitcher, up glass, stirrer if up
GARNISH   None

Godmother

Level   3
MAJOR INGREDIENTS   Amaretto, vodka
GLASSWARE   Same as for Godfather
GARNISH   None

Rusty Nail

Level   3
MAJOR INGREDIENTS   Scotch, Drambuie
GLASSWARE   Same as for Godfather
GARNISH   None

Jack Rose

Level 2
MAJOR INGREDIENTS   Apple Jack (apple-flavored brandy), sour mix, grenadine
ASK   "Up or on the rocks?"
GLASSWARE Same as for sour
GARNISH   None

Collins

Level 2
MAJOR INGREDIENTS   Vodka, club soda
  Tom Collins   Gin, club soda
GLASSWARE   Red wineglass or highball
GARNISH   Orange, cherry, splash of soda

Sloe Gin Fizz

Level 2
MAJOR INGREDIENTS   Sloe gin, sour mix
GLASSWARE   Red wineglass or highball
GARNISH   Orange, cherry, splash of soda

Singapore Sling

Level 3
MAJOR INGREDIENTS   Gin, grenadine, orange juice, sour mix, Cherry Herring, club soda
GLASSWARE   Redwine or highball
GARNISH   Club soda

Piña Colada

Level 4
MAJOR INGREDIENTS   Rum, pineapple juice, coconut milk
GLASSWARE   Champagne or specialty drink
GARNISH   Pineapple slice, straw

Planter's Punch

Level Misc Liquor - $5.00 (or whatever the house charges)
MAJOR INGREDIENTS   Myers Rum, grenadine, orange juice, pineapple juice
GLASSWARE   Iced tea or specialty drink
GARNISH   Orange slice, strawberry, pineapple, lime on rim of glass; cherry on top; straw

Mai Tai

Level Misc Liquor - $5.00 (or whatever the house charges)
MAJOR INGREDIENTS   Rums, fruit juices
GLASSWARE   Iced tea or specialty drink
GARNISH   Same as for Planter's Punch

Zombie

Level   Misc Liquor - $6.00 (or whatever the house charges)
MAJOR INGREDIENTS   Various rums, fruit juices
GLASSWARE   Iced tea or specialty drink
GARNISH   Same as for Planter's Punch

Scarlett O'Hara

Level   2
MAJOR INGREDIENTS   Southern Comfort, cranberry juice
GLASSWARE   Highball
GARNISH   None

POPULAR VODKA DRINKS

Vodka Tonic

Level   1
MAJOR INGREDIENTS   Vodka, tonic water
                    Gin, tonic water if a "Gin Tonic"
GLASSWARE   Highball
GARNISH   Lime on the rim of glass

Bloody Mary

Level   1
MAJOR INGREDIENTS   Vodka, Bloody Mary mix
GLASSWARE   Red wineglass or highball
GARNISH   Celery stick, Lime on rim of glass
VARIATIONS   Screwdriver: Vodka, Orange juice
             Greyhound: Vodka, Grapefruit juice
             Salty Dog: Vodka, Grapefruit juice, salt rim
             Sea Breeze: Vodka, Grapefruit, cranberry juices
             Bay Breeze: Vodka, Pineapple, cranberry juices
             Cape Codder: Vodka, Cranberry juice
             Madras: Vodka, Cranberry, orange juices
             Bloody Bull: Vodka, Bloody Mary mix, bouillon
             Bull Shot: Vodka, Bouillon, Tabasco, worcestershire sauce

Gimlet

Level   2
MAJOR INGREDIENTS   Vodka, Rose's Lime Juice; gin, Rose's Lime Juice
ASK   "Up or on the rocks?"
      "Vodka or gin?" (unless they specify)
NOTE   Add a "call" or "premium" surcharge (e.g., 50 cents for "call"; $1.00
       for "premium") (e.g., "Beefeater's") if customer requests brand.
GLASSWARE   Rocks or pitcher, up glass, stirrer
GARNISH   Lime (on rim)

Harvey Wallbanger

Level   3
MAJOR INGREDIENTS   Vodka, Galliano, orange juice
GLASSWARE   Highball
GARNISH   None

Melon Ball

Level   2
MAJOR INGREDIENTS   Madori, vodka, orange juice
GLASSWARE   Highball
GARNISH   None

Stinger

Level   3
MAJOR INGREDIENTS   Vodka, white crème de menthe; scotch, white crème de menthe for a Scotch Stinger
GLASSWARE   Same as for Godfather (usually rocks)
GARNISH   None

Tequila Sunrise

Level   3
MAJOR INGREDIENTS   Tequila, orange juice, grenadine
GLASSWARE   Highball
GARNISH   None

CREAM DRINKS

House policy question: Are ice cream drinks available? If so, they are usually served "up," strained through ice in the mixing pitcher into an "up" glass.

Toasted Almond

Level   4
MAJOR INGREDIENTS   Amaretto, Kahlua, cream
ASK   "Up or on the rocks?" (usually "up")
GLASSWARE   Rocks or pitcher, up glass, clear plastic stirrer

Kahlua & Cream

Level   2
MAJOR INGREDIENTS   Kahlua, cream
ASK   "Up or on the rocks?" (usually "up")
GLASSWARE   Rocks, or pitcher, up glass, stirrer
GARNISH   None

Black Russian

Level   3
MAJOR INGREDIENTS   Kahlua, vodka
NOTE   All else same as for Kahlua & Cream.

White Russian

Level 3
MAJOR INGREDIENTS   Kahlua, vodka, cream
NOTE All else same as for Kahlua & Cream.

Brandy Alexander

Level 4
MAJOR INGREDIENTS   Brandy, crème de cacao, cream
ASK   "Up or on the rocks?"
GLASSWARE   Rocks, or pitcher, up glass, stirrer
GARNISH   Nutmeg

Pink Squirrel

Level 3
MAJOR INGREDIENTS   Crème de noyaux, white crème de cacao, cream
ASK   "Up or on the rocks?"
GLASSWARE   Rocks, or pitcher, up glass, stirrer
GARNISH   None

Pink Lady

Level 3
MAJOR INGREDIENTS   Gin, cream, grenadine
ASK   "Up or on the rocks?"
GLASSWARE   Rocks, or pitcher, up glass, stirrer
GARNISH   None

Grasshopper

Level 3
MAJOR INGREDIENTS   Green crème de menthe, white crème de cacao, cream
ASK   "Up or on the rocks?" (usually up)
GLASSWARE   Rocks, or pitcher, up glass, stirrer
GARNISH   None

Golden Cadillac

Level 3
MAJOR INGREDIENTS   Galliano, white crème de cacao, cream
ASK   "Up or on the rocks?" (usually up)
GLASSWARE   Rocks, or pitcher, up glass, stirrer
GARNISH   None

Golden Dream

Level 3
MAJOR INGREDIENTS   Galliano, Triple Sec, vodka, orange juice, cream
ASK   "Up or on the rocks?" (usually up)
GLASSWARE   Rocks, or pitcher, up glass, stirrer
GARNISH   None

CHAMPAGNE DRINKS

Serve these drinks in champagne tulip glassware.

Kir Royale

Level 3
MAJOR INGREDIENTS Champagne, Cassis
GARNISH Twist

Champagne Cocktail

Level 3
MAJOR INGREDIENTS Champagne, bitters, sugar
GARNISH Twist

Mimosa

Level 3
MAJOR INGREDIENTS Champagne, orange juice
GARNISH Orange slice

Poinsettia

Level 3
MAJOR INGREDIENTS Champagne, cranberry juice
GARNISH None

APERITIFS

Campari

Level 1
MAJOR INGREDIENTS Campari, soda (default) or tonic or orange juice
GLASSWARE Highball
GARNISH Twist or lime. Ask which is preferred.

Dubonnet

Level 1
MAJOR INGREDIENTS Blond (white) or Rouge (red) Dubonnet
GLASSWARE Rocks or pitcher, up glass, stirrer
GARNISH Twist. Optional; ask

Lillet

Level 1
MAJOR INGREDIENTS Blond (white) or Rouge (red) Lillet
(Often served as a spritzer, with soda)
GLASSWARE Rocks or pitcher, up glass, stirrer; red wineglass if served as a spritzer
GARNISH Orange slice with white lillet, optional

Vermouth

Level 1
MAJOR INGREDIENTS   Dry or sweet vermouth
GLASSWARE   Rocks or pitcher, up glass, stirrer (usually served on the rocks)
GARNISH   None

Sherry

Level 2
MAJOR INGREDIENTS   Sweet sherry: Harvey's Bristol Cream
Semisweet: Dry Sack Sherry
Dry: Tio Pepe Sherry
GLASSWARE   Usually a cordial glass; occasionally, rocks
GARNISH   Twist (optional)

Port

Level 3
GLASSWARE   Usually a cordial glass; occasionally, rocks
GARNISH   None

CORDIALS

Level 2
Kahlua
Amaretto
Drambuie*
Crème de Menthe (green, white)
Crème de Cacao (light, dark)
Ouzo
Benedictine*
Anisette
Tia Maria
Peppermint Schnapps
Peach Schnapps
Blackberry Brandy
Peach Brandy
Galliano
Sambuca (Gets 3 coffee beans if served up)
Strega
NOTES Serve cordials in cordial glasses if up;
Serve cordials in rocks glasses if on the rocks.

Level 3
Frangelico
Framboise
Chambord
Chartreuse (green, yellow)
Grand Marnier*
Bailey's Irish Cream
B&B*

COGNACS

Serve in brandy snifters. If asked to warm it, run hot water through the snifter then dry it. Do not flame brandy or rum inside a brandy snifter to warm the glass—the glass might break.

Level 2	*Level 4*	*Misc Liq* ($5.00)	*Misc Liq* ($10.50)
House Cognac	Rémy Martin	Martell Cordon Blue	Martell Cordon Blue
Courvoisier	Armagnac	Henessey V.S.	Xtra
Pear Brandy	Calvados	Hine Cognac	
	Williams Pear Brandy		

COFFEE ALCOHOLIC DRINKS

General Procedure

1. Get tray, coffee pot, fresh hot coffee, small dish of whipped cream from the kitchen.
2. Enter the liquor order into the point-of-sale system if there is one and obtain the beverage dupe.
3. Proceed to the bar and exchange the beverage dupe for liquor. Use a white wine or specialty drink glass or whatever the operation specifies.
4. Once you have liquor, proceed to the station. Don't go into the kitchen with liquor, especially in a glass. (Glass could break in kitchen area.)
5. Assemble the coffee/liquor drink at the side station:
   Add 1 teaspoon sugar to the liquor in the glass.
   Stir.
   Add hot coffee to liquor in glass.
   Spoon whipped cream over coffee.
   Place stirrer in glass.
   Serve on B&B plate with doily or follow operation policy.

PRICE	Most are *Level 4*
GLASSWARE	White wine or specialty drink glass
GARNISH	Whipped cream, stirrer
TYPES	Irish Jamison's Irish Whiskey
	Italian Amaretto
	Sambuca
	Bailey's Irish Cream
	Jamaican Tia Maria
	Kioki Cognac, Kahlua
	Mexican Kahlua
ASK	"Would you like it sweetened a bit?"

Some people want only coffee, liquor, whipped cream; most people expect a bit of sugar, liquor, coffee, whipped cream.

Default: Add 1 teaspoon bar sugar to liquor at the bar.

*Aromatic liqueurs may be served in either brandy snifters or cordial glasses.

BASIC LIQUORS

Bar or Well Liquor (usually Level 1)

Bourbon	Scotch	Rye
Blended whiskey	Vodka	Gin

Call Liquors (usually Level 2)

Rums:	Myers (Dark)	
	Bacardi (Light, Dark, Gold Reserve)	
	Mt. Gay (Dark)	
Tequila:	Cuervo Gold	
Gin:	Beefeater's	
	Gordon's	
	Tanqueray	
	Bootles	
	Bombay	
Vodka:	Stolychnaya (Stoly)	
	Finlandia	
	Absolut	
Rye:	Canadian Club (CC)	
	Seagram's 7	
	Seagram's V.O.	
Bourbon:	Wild Turkey	
	Old Grand Dad	
	Jack Daniels	
Blended Scotch:	Cutty Sark	
	Chivas Regal	(Level 3)
	Crown Royal	
	Pims	
	Johnny Walker Red	
	Johnny Walker Black	(Level 3)
Unblended Scotch:	Glenlivit	(Level 3)
	Glenfiddich	(Level 3)
Irish Whiskey:	Jameson's	
	Old Bushmills	
	Murphy's	

PROCEDURE FOR SODA SERVICE

Club Service

1. Place the glass with ice in it on the table, to the right of the guest, using your right hand. Place the straw, wrapped, to the right of the guest.
2. Remove the bottle of soda from the beverage tray and pour soda from the bottle into the glass.
3. Place the bottle, with the remainder of the soda in it, on the table for the guest to finish pouring at leisure.

Fountain Service

1. At the service bar, place ice in the glass.
2. If there is a soda gun, select the proper button on the gun and fill to within one-half inch of the rim of the glass.
3. If there is not a soda gun, fill the glass with the guest's choice of soda using the method prescribed by the establishment.

Juices

Serve juices in a white wineglass or a juice glass if the guest desires the juice up. Serve juices in a red wine or highball glass if the guest desires the juice on the rocks.

Know in advance whether your establishment is serving fresh-squeezed or prepared juices. The juice can be labeled as "fresh-squeezed" if it has been prepared from fresh fruit within the past 24 hours. If the juice was prepared from frozen concentrate or came canned, frozen, or bottled in a carton, it cannot be sold as "fresh-squeezed."

The following juices are popular either alone or as mixers with other ingredients:

Orange	Preferably squeezed fresh daily
Grapefruit	Sometimes fresh; usually bottled or canned
Cranberry	Usually bottled or canned
Apple	Usually bottled or canned

Drinks using the preceding juices and various other fruit and vegetable juices are detailed in Chapter Eight.

Waters

Various bottled waters and tap water are being requested by more and more guests. Water should be served in either a highball glass or a red or white wineglass, with ice. Ask the guest if lemon or lime is desired with the water and serve with a lemon or lime wedge or slice, depending on the house preference. Some establishments automatically serve water with lemon or lime, as a house standard.

Chapter Eight provides a detailed listing and discussion of bottled waters.

Wine Coolers

Wine coolers are similar to spritzers. They should be served in a red wineglass, with ice. Often they are garnished with a twist, but ask the guest first whether a garnish is desired.

Spritzers

Spritzers generally are served in a highball or red wineglass. Usually, a spritzer consists of the house red or white wine over ice, finished with club soda and a twist. If a guest requests a Lillet spritzer, use the white Lillet and an orange slice, not a twist, as the garnish.

Kirs

A Kir is a white wine with a dash of crème de cassis (black currant liqueur), served either up or on the rocks. If served up, a white wine glass is used; if served on the rocks, a red wine or highball glass is used. The garnish is a twist.

A Kir Royale is made with the house champagne and a touch of crème de cassis and is served up, in a fluted champagne glass. There is no garnish.

Serving Brandy

Brandy should be served from the bar in a balloon-shaped glass on a short stem called a "snifter" or "inhaler." Another, less popular option is to serve it in a small, slightly rounded, elongated glass on a tiny stem, called a "pony." The server carries the poured brandy to the customer and places it above the coffee cup. If there is no coffee cup, the brandy is placed to the right of the guest where a coffee cup would normally be placed.

PROCEDURE FOR POURING BRANDY TABLESIDE

1. Present the bottle: "Your 'Martell Extra,' sir/madam."
2. Remove the stopper from the bottle.
3. Place the glass to be poured into at the edge of the table.
4. Pour 1 ounce (or the house pour) into the glass.
5. Place the glass above the coffee cup.

Use of a snifter allows the drinker to warm the brandy by cupping the brandy snifter in the palm of the hand, using body heat to warm the contents. With that bit of heat applied, brandy releases its delicate bouquet and can be swirled around against the inside of the snifter glass, intensifying the aroma.

Brandy may also be poured at the table in much the same way wine is poured. The bottle is presented first, especially if it is a superior brandy, then poured.

Sometimes an operation will have an elaborate ceremony for pouring brandy that includes warming the glass in a special device. This is not a usual practice and requires inventorying cumbersome equipment to be utilized only infrequently. Warming of the brandy by the placement of the guest's palm on the bowl of the snifter is not only sufficient, it is preferred.

Serving Beer and Other Malt Beverages

When the beer is on "draught," dispensed from a faucet behind the bar, the server has only to hurry it to the customer when it is drawn. When it is served in bottles, the server should open the bottle at the bar and pour the beer, or at least begin to pour, for the customer. In the United States, customers usually prefer beer ice-cold, and it should be served that way. The guest who prefers the beer "European style," will want it slightly warmer (however, it will warm up just from being in most rooms).

Domestic or Import?

Beer is usually categorized as either imported or domestic. Often, the prices are different for each category with the imported beer being slightly more expensive. The increase in price does not necessarily indicate a corresponding increase in quality but more probably reflects import taxes and increased shipping charges.

PROCEDURE FOR SERVING BEER

1. If possible, use chilled (frosted) pilsner glasses or mugs for beer. If there is not sufficient space to keep a freezer for "beerware," rinse the glass or mug in cold clean water before approaching the table. (All beer glasses should be rinsed carefully after being washed because any soap film or residue will cause the beer to go flat.)

2. If the bartender does not open the bottle of beer, open the bottle either at an affixed bottle-cap remover at the bar area or with a "church key," bottle opener, or even corkscrew. If using a corkscrew, pry the cap off the bottle with the lifter that is part of a corkscrew. Don't flip the cap (where will it go?), but lift it off then catch the cap in your hand. Remove the corkscrew from the bottle and dispose of the cap in a garbage container.

3. When you reach the customer, ask whether he or she would like the beer poured. If the guest says, "No," place the chilled glass to the guest's right, and the bottle of beer to the right of the glass.

4. If the guest says, "Yes," first place the chilled glass to the guest's right, if possible. If using a pilsner glass, hold the bottle directly over the center of the glass, with the rim of the bottle lip near one side of the glass. The glass should have sloping sides, with a small base and wide rim at the top to provide a sloped pouring surface.

5. If a large head is desired, pour directly into the center. Otherwise, the beer should be poured down the side of the glass to minimize the head. Those who are adept at pouring will be able to create a head that rises 3/4 inch over the top of the glass and settles without spilling. The head should account for no more than 20 percent of the filled glass.

6. Place the bottle, with the remainder of the beer, on a coaster or beverage napkin.

7. Some guests will ask you to pour, "Just a bit." In this case, begin to pour the beer but stop after pouring 2 or 3 ounces, or when the glass is approximately one-third full. The guest can then refresh the glass as desired.

DOMESTIC AND IMPORTED BEERS

Domestic	Imported	Country of Origin
Ballentine	Affligem	Belgium
Black Label	Amstel Light	Holland
Blatz	Augustijn	Belgium
Budweiser	Bass Ale	United Kingdom
Bud Light	Beck's	West Germany
Busch	Calgary Amber Lager	Canada
Colt 45	Chihuahua	Mexico
Coors	Corona Extra	Mexico
Coors Light	Dos Equis	Mexico
Dixie	De Koninck	Belgium
Gennessee	Dragon Stout	Jamaica
I.C.Light	Dribeck Light	West Germany
Iron City	Duvel	Belgium
Lone Star	Foster's	Australia
Michelob	Great Wall	China
Milwaukee's Best	Grizzly	Canada
Miller	Grolsch	Holland
Miller Lite	Guinness Stout	England
National Beer	Heineken	Netherlands
New Amsterdam	Kaliber	Ireland
Old Milwaukee	Kirin	Japan
Ortleib	Kwak	Belgium
Pabst Blue Ribbon	Molson Ale	Canada
Rheingold	Moosehead	Canada
Rolling Rock	Mort Subite	Belgium
Samuel Adams	Moussy	Switzerland
Schaeffer	O'Keefe	Canada
Schlitz	Pacifico	Mexico
Schmidt's	Palm	Belgium
Stony	Polar Beer	Venezuela
	Red Stripe	Jamaica
	Rodenbach	Belgium
	Scaldis	Belgium
	Swan Lager	Australia
	Tecate	Mexico
	Tsingtao	China
	Wittekop	Belgium

The various brands of beer served by the operation and the price/ category for each beer should be memorized by each server and bartender.

Glassware Overview

Certain glasses are usually designated by management to be used only for certain beverages. By tracking the history of drinks served, it is possible to forecast which drinks are ordered frequently, less frequently, and rarely. With this information, management, in turn, can forecast how many of each type of glass will be needed during a slow shift, an average shift, a very busy shift, and a holiday shift.

Par levels can then be established for each type of glass, including the number of glasses required to be in service for a given shift and the number needed for backup (glasses boxed and not in service but available if needed). The service staff thus can be confident that there will be sufficient glassware for the service period. It is usually a duty of the bar manager to ensure that sufficient glasswares are available for the upcoming shift by checking the number of each type of glass in the bar available for use, the number of each in the warewashing area, clean but not yet moved to the bar area, and those still to be washed. The bar manager generally adds an estimate for the number of each type of glass in use in the dining room depending on the time of day and day of the week he or she is checking the inventory.

Specialty glassware is often used for house drinks and for rare wine service. Specialty drinks often are served in oversized stemmed wine goblets that are topheavy and prone to breakage. The price of the glassware and breakage needs to be calculated into the cost of the drink.

Likewise, when pricing the rare wines, the price, storage, care, and breakage factors for rare wine glassware need to be taken into account. Rare wines should be served in stemmed, clear, thin leaded crystal glasses with the proper shape for the type of wine being served (see Chapter Eleven for examples of traditional styles of wineglasses). Glasses used for the drinking of rare wines should not have a rolled edge or lip as this detracts from experiencing the rare wines. These thin, stemmed, leaded crystal glasses not only are expensive to buy but must be stored apart from the everyday glassware and must be hand washed and dried. Further, unless used often, they must be hand steamed and hand dried before each use to ensure that no dust or other particles have collected on or in the glass that would (1) detract from the clarity of the glass, perhaps mistakenly giving the wine a cloudy color or (2) detract from the taste of the wine, perhaps giving it an "off-taste." Either of these scenarios might occur if there is dust on the outside of the bowl or inside the bowl of the glass.

The chart identifies the glassware typically used for common alcoholic drinks. A similar chart should be available for all service personnel. Those involved in ordering, picking up, serving, and making drinks need to memorize the information.

GLASSWARE AND GARNISHES FOR ALCOHOLIC BEVERAGES

Glass	*Beverage*	*Garnish*
Rock (short, squat)	Martini, Rx	Twist or olive
	Gibson, Rx	Onion
	Manhattan, Rx	Cherry
	Straight liquor, Rx	Twist
	Straight liquor, Up	
	Sherry, Rx	
	Old Fashioned	Sugar, bitters, cherry, orange slice
	Cordials, Rx	
	Dubonnet, Rx	
	White Russian	
	Black Russian	
	Sombrero	
Highball (tall, straight sides)	Gin & Tonic	Lime wedge
	Rum & Coke	
	CC & 7	
	Scotch & water	
	Bourbon & soda	
Cordial (small, Y-shaped)	Sambuca	
	Crème de menthe	
	Sherry, Up	
	Port, Up	
	Shots of liquor	
White Wine (straight sides, tall, narrow; stem)	White wine	
	Irish coffee	Whipped cream
	Juices, Up	
	Perrier	Lime wedge
	Bloody Mary, Up	Lime, celery stick
	Sours	Cherry, orange slice
	Collins, Up	Cherry, orange slice
	Marguerita, Rx	Salt rim, lime
	Brandy Alexander, Rx	Nutmeg
	Lillet	Orange slice
	Hot buttered rum	
	Regular Daiquiri	Lime

Glass	*Beverage*	*Garnish*
Red Wine (wide diameter bowl; stem)	Red wine	
	Juices, Rx	
	Bloody Mary, Rx	Celery, lime
	Sodas	
	Spritzer	
	Kir, Rx	Twist
	Collins, Rx	Cherry, orange slice
	Double highball	
	Sangria	
	Milk	
Champagne Tulip (tall, curved sides; stem)	Champagne	
	Strawberry Daiquiri	Strawberry, Sugar rim
	Piña Colada	Pineapple
	Champagne cocktail	Sugar, bitters
	Mimosa	Orange slice
Brandy Snifter	Brandy	
	Cognac	
	Benedictine	
	B&B	
	Double Bloody Mary	Celery stick, lime
Cocktail or Up Glass (w/pitcher)	Martini, Up	Ol or Tw
	Manhattan, Up	Cherry
	Gibson, Up	Onion
	Grasshopper, Up	
	Cream drinks, Up	
	Marguerita, Up	Salted rim
Beer Pilsner Glass	All beers	

Rx = "on the rocks" or with ice
Up = without ice
Tw = lemon twist
Ol = olive

SAMPLE PRICE LEVELS FOR COCKTAILS

Liquor	Level 1	Level 2	Level 3
Gin	Gin & Tonic Martini Gibson	Tanqueray Bombay Beefeater's Boodles Gordon's Manhattan Gin Gimlet Tom Collins	Singapore Sling
Vodka	Vodka Martini Vodka Gibson Screwdriver Bloody Mary Greyhound Madras Cape Cod	Stoly Absolut Finlandia Smirnoff's Gimlet Collins Melon Ball	Godmother Stinger Harvey Wallbanger Golden Dream
Bourbon		Jack Daniels Wild Turkey Old Grand Dad	
Scotch	Scotch/Water Scotch/Soda Scotch Sour	Dewars JW Red Label Jack Daniel's Cutty Sark Crown Royal Pims Rob Roy	JW Black Label Chivas Regal Glenlivit GlenFiddich Godfather Rusty Nail
Rye		Seagram's 7 Seagram's V.O. Canadian Club Old Fashioned 7 & 7	
Rum	Rum & Coke	Bacardi Myers Mt. Gay	Daiquiri (reg)
Tequila		Cuervo Gold Sauza	Marguerita Sunrise
Others	Dubonnet Campari Lillet Vermouth	Kahlua Amaretto Schnapps Ouzo Anisette Galliano	Pink Squirrel Pink Lady Grasshopper Golden Cadillac Golden Dream Frangelico Framboise

❖ BASIC WINE SERVICE

If alcoholic beverages, especially wines, are served in an establishment two positions are of paramount importance—the bartender(s) for mixology and cocktails, and the wine sommelier (wine steward), or other responsible person, for knowledge about the house wine list. In addition, everyone who serves alcoholic beverages and all managers must be aware of legal considerations concerning the service of alcoholic beverages.

The wine-knowledgeable person in an upscale restaurant is often a sommelier or maître d'hôtel. Otherwise, it is usually a captain or server, or it might be all captains and/or servers on the staff if the establishment has regular staff wine information meetings and tastings. Whoever the wine-knowledgeable person(s) may be, knowledge of wines, food, and how to sell both wines and food are the basic skills necessary.

The sommelier, or wine steward, is usually the most knowledgeable person in the operation regarding wines and is usually also in charge of the following:

♦ Purchasing wines, or recommending them for purchase.
♦ Proper storage of wines (temperature, humidity, and physical placement of wine bottles on their sides).
♦ Decanting of wines, if necessary.
♦ Wine lectures and tastings for the staff.
♦ Wine handouts for the staff.
♦ Sale and upselling of wines.
♦ Opening and pouring of rare wines.
♦ Tasting of customer-rejected wines, especially if the wine is suspected of having a "bad" or "off-odor."
♦ Inventory control.

If a sommelier is not present, a manager or food and beverage (F&B) director assumes most of the preceding duties. The service staff, captains and servers, often handle the normal wine sales and service—selling, upselling, opening and pouring of wines on the dining room floor. Managers and sommeliers are generally called only when there is a difficult cork or a rare wine to open, or when there is a problem with a wine's smell or taste.

This section reviews wines—their background, how they are made, general types of wines, and how to open, pour, and serve most wines.

Few people dine without also drinking some beverage. The most commonly served beverages are water, then sodas, then coffee.

For an elegant or leisurely meal, however, wine is increasingly being ordered.

Since this is so, service staff should be knowledgeable in all aspects of wine service. Professional wine service demands skill. This section deals with the techniques needed to be a professional wine server and salesperson. The more a person knows about wines—how each tastes, which wine "flavors" complement which foods—the easier it is to sell wines to guests. Guests who want wine with their meal but cannot decide which wine, or are afraid to choose one, will be pleased if a knowledgeable server makes an appropriate suggestion, and they will order the wine. The guests will be elated, the server quite pleased, the management happy, and the operation more profitable. Having a service staff knowledgeable in wines and how to recommend them, how to pair wines with food, and how to provide "flair" while opening and serving them, can be a real difference maker.

Essential Skills for Serving Wine

Each stage in the service of wine requires a particular skill:

♦ Presenting the wine list.
♦ Setting the glasses.
♦ Presenting the bottle.
♦ Opening the bottle.
♦ Serving the wine.

The following sections discuss these skills.

Presenting the Wine List

The wine list (if a restaurant has a separate one) is presented from the customer's left. It should be presented at the same time as the food menu. Usually one wine list per table is presented; if there are six or more guests, however, two or more wine lists may be presented. Presenting two or more wine lists encourages discussion about which wine to choose, with opposing sides of the table rooting for their favorite wine. Perhaps two or more bottles will be ordered.

Sometimes an operation's wine list is printed on the main menu. If no list is apparent, the customer may ask the server to describe the available wines or bring a wine list to the table. If the operation sells wines, the sommelier, captains, and/or servers should know:

♦ The types of wines available and their names, with correct pronunciation.
♦ The country of origin of each bottle offered.
♦ The color of the wine.
♦ The style of the wine—dry, sweet, medium-bodied, and so on.

When wine is offered, the captain, sommelier, and/or server will often be asked to discuss the wine list with guests and to make recommendations. Knowledge of the wines on the list as well as wines in general is necessary to relate the wines being offered to those the customer is familiar with drinking. During service periods, someone should be available at all times who is knowledgeable regarding the wines the operation offers, which foods on the menu each complements, and each wine's price. This information is essential for all personnel who have guest contact.

If a guest asks for suggestions, ask first if the person prefers red or white wines and dry, medium, or sweetish wines. This narrows the list to the color preference and general level of sweetness (or residual sugar) the guest prefers. Then suggest three wines of that color and residual sugar level from different countries (e.g., France, the United States, and Italy), if there are appropriate wines from three different countries on the wine list. Suggest wines in different price ranges—an inexpensive, a moderate, and a higher-priced wine.

Guests may have a "country" preference (e.g., French wines), or a price range for wine expenditures. Unfortunately, guests with such preferences or restrictions often don't share them with the service staff, but the preceding technique allows the guest to choose a wine within his or her preference without seeming obvious.

For instance, if a guest prefers American wines and three French wines are suggested, the guest might assume that none of the American wines are good or that the restaurant is "pushing" French wines. Likewise, if all the selections are expensive, the guest might assume that the server is only trying to increase the check for a larger tip. The server may have, with that one suggestion, lost credibility with the guest.

If wines are being merchandised (featured on tabletop tents, or in a special display case), the service staff must be especially knowledgeable about those wines. They can't sell a product unless they are aware of its positive attributes.

Some operations have sales contests whereby the sommelier, captains, and/or servers receive prizes or a payment for each bottle sold ("bottle bonuses"). To augment these sales through suggestion, there may be a bottle or carafe of wine on each table. As a means of tabulating sales and determining a contest winner, POS daily reports can track sales, or servers/captains may save the corks from the bottles they sell. If a guest takes a cork home, a copy of the guest check usually suffices as proof of sale. At the end of a stipulated period (usually a week or a month), the winners are announced and the prizes awarded. Wineries or distributors often sponsor such contests to increase a particular wine's sales.

When the customer has made a choice, with or without service staff assistance, the server/captain should pronounce the name of the selected wine, tell the customer its bin number (if one is shown on the wine list) and mention the year, if it is a vintage wine. Note whether a full or half bottle is being ordered and vintage year, if any. The price is not repeated.

The server/captain then obtains the wine from the storage area (in the dining room, from the bar, or from the wine cellar) or a bar dupe or cellar check is sent to the wine cellar indicating the wine needed (for example bin number 21) and whether a bottle or a half-bottle (bin number 19 × $\frac{1}{2}$) is being ordered.

Setting the Glasses

In upscale restaurants, a wineglass may already be set on the table. A clear, stemmed wineglass on the table lends an air of formality and adds height to the table setting. It also suggests to the guests that they ought at least to be thinking about ordering wine if wine has not been ordered already.

The preset, all-purpose glass may be the wrong glass for the wine ordered. For example, the customer may have ordered champagne, which should be poured into a tall flute and not into a glass suitable for still wine. In this case, the proper glasses are placed, and the wineglasses that will not be used are removed from the table.

Should several wines be ordered (champagne for a toast, white wine for the appetizer, red wine for the entrée), the additional glasses that will be needed are brought to the table and added to the place settings. In this case, the all-purpose wineglasses could be used for either the white or red wine; additional glasses would be brought for the other wine and champagne flutes would be provided for the toast.

Bussers, servers, captains, and sommeliers must know which glasses the operation uses for particular drinks, including wines. Many restaurants purchase multiple sizes, shapes, and designs of glassware. Often, consistency with the tableware and the theme of the restaurant is the first consideration. In many operations, an all-purpose wineglass is used for all wines, red or white, with a

STANDARD WINEGLASSES
(See p. 214 in Chapter Eleven.)

1. All-purpose wineglasses are generally stemmed, clear, and tulip-shaped. They hold between 6 and 8 ounces of wine.
2. Rhine wineglasses are tall and stemmed, with squat cylindrical bowls. They hold about 6 ounces.
3. White wineglasses are clear, stemmed, and tulip-shaped but hold less than red wineglasses—about 6 ounces.
4. Red wineglasses are generally the operation's largest tulip-shaped glasses, holding 8 to 12 ounces. They also should be clear and stemmed.
5. Champagne glasses, also used for other sparkling wines, are either elongated tulips or saucer-shaped, shallow glasses on long stems. The flute shape (elongated tulip) is the preferred glass shape for sparkling wines because the smaller surface area exposed to air keeps the bubbles in the wine for a longer period. In addition, with the flute shape, guests can enjoy watching the bubbles rise to the top of the glass from the bottom.

champagne flute being the only other glass used for wine service. This eliminates the question of which wineglass to use.

Generally, glasses are set above the right-hand side of the cover. If one wineglass is used, the point of the main course knife should be in exact line with the stem of the glass. Other glasses, when used, are often arranged to the right in a straight line from this glass. Usually, they are arranged in the order of service; the first glass to be used is at the extreme right, and the guest works upward and to the left with each new wine.

There are alternative acceptable settings. The glasses might be arranged by size, with the smallest on the far right, then progressing upward and leftward. When three glasses are set, the glasses can be arranged in a triangle—one glass above the knife, one glass immediately to the right of it, and one glass below the second glass, between the two top glasses. Often, a water glass will be placed to the left of the tip of the entrée knife, with the wineglasses at the tip of the entrée knife and to the right.

When handling glassware for wine service, remember these rules:

1. In putting glasses on the table, place them; do not slide glasses.
2. Handle all stemware glasses by the stems, not by the globe.

3. If using glasses without stems (tumbler-style wineglasses), handle them by the base (bottom 2 inches) of the glass.

The motif of the restaurant determines whether wines are served in stemware. Some restaurants that pride themselves on a casual, down-home atmosphere also pride themselves on using jug wines and tumblers for wine (or water or milk) service. Establishments that are located next to or on water may use plastic "glassware" (perhaps even tumbler-style) for breakage reasons and for the safety of their guests (many of whom might be barefoot or wearing sandals).

Many guests and connoisseurs prefer stemware because it allows the wine to remain at the proper temperature and release its bouquet. The beverage is not warmed by the heat of the hand, and the globe shape allows the wine to swirl easily, releasing its bouquet. However, if guests are quaffing the wine and don't care about evaluating a wine's color, bouquet, legs, or body, it really doesn't matter much which style of glass is used. Plan to use the glass that fits into the decor, the motif, and the theme of the restaurant or establishment, whether those glasses are stemmed or not, crystal or not, safety-rimmed or not.

Presenting the Bottle

Before a bottle of wine is opened, it should be presented to the guest who ordered it. The presentation serves the following purposes:

1. The wrong bottle of wine may have inadvertently been brought to the guest.
2. The vintage of the bottle may not agree with the vintage printed on the wine list.
3. It provides a "last opportunity" for the guest to see the wine ordered before it is opened. At many restaurants, the point at which the bottle is opened is when it has officially been purchased unless the guest and someone in management deem the wine is "bad."

Bad wines are usually determined to be so, or not so, by a manager, maître d', or sommelier, as well as by the guest. In the interest of promoting guest satisfaction and having guests return, however, most operations will agree with the guest whether or not the bottle seems bad to the management.

Sometimes a bottle of wine with an unusual odor will not be "bad" but merely needs air time. There have been several instances whereby wine was found to have "bad" corks and the wine smelled

---

PROCEDURE FOR PRESENTING
BOTTLES OF WINE

1. Approach the guest from his or her right-hand side.
2. Hold the base of the bottle in the palm of the left hand. Have a folded, clean napkin immediately beneath the bottle. Hold the neck of the bottle with the right hand. Position the bottle so the label is facing the guest.
3. Tilt the bottle so that the guest can easily read the label.
4. Handle the bottle carefully, without making sudden moves. Do not shake the wine. There may be sediment in the bottle, especially if it's an old red wine, that will be distributed throughout the wine if shaken and that will take days to settle back to the bottom.
5. Announce the name of the wine and the vintage (year), if any; for example, "Your Chateau Issan '53."

"corky." After allowing the bottle to breathe for an hour or so, it was drinkable. If the bottle is, in fact, fine and drinkable, most establishments will sell it by the glass at the bar, substituting it for the house wine for the next six or so orders of house wine or offering it as a special "wine by the glass."

Opening the Bottle

After presenting the wine, proceed to open it.

Still Wines

In many European or European-style restaurants, it is house policy for the server to leave the table after presenting the wine, opening it in the pantry or kitchen or bar area. In most U.S. operations, however, wine is opened at the table.

Red wines are opened either on the tabletop, with the label facing the host, or in the sommerlier's hands, at the side of the guest who ordered the wine.

White wines, rosé wines, and chilled Beaujolais wines are opened in an ice bucket that should have been brought to the table, filled two-thirds full with ice water or salted ice water to chill the wine quickly. If a white wine is being opened in the pantry, the wine is returned to the table with the cooler and stand.

Sparkling Wines

Champagne and other sparkling wines are sealed with mushroom-shaped corks that protrude from the bottle and are, in turn, covered

TECHNIQUE FOR OPENING STILL WINES

Equipment Needed

- Clean service cloth, folded into quarters along the length, then draped over the left arm during the opening ceremony. Inside corners of this cloth will be used to wipe the lip of the bottle while removing the cork and immediately after removing the cork and for catching "drips" between pours.
- Server's corkscrew, with a straight-edge knife and 2-inch worm.
- Ice bucket for white, rosé, and Beaujolais wines.
- Tasting glass (optional).

1. For screwtop bottles, use the knife portion of a waiter's corkscrew to slit the plastic sleeve around the neck and remove it. Unscrew the cap.

   For wines with corks, use the knife portion of the corkscrew to cut the tinfoil or plastic capsule around the neck and lip of the bottle. Cut this below the second (or lower) lip so about 1/2 inch of bottle is showing. Cut across the top of the foil and remove it.

   This procedure should be done in two smooth motions, without turning the bottle, which might shake any sediment. If it's an aged red wine, the guest might refuse the wine if sediment is poured with the wine. One motion is utilized to turn the hand and corkscrew around the entire lip of the bottle. The second motion is to take the knife up, across, and down the side of the foil cap, cutting it into two pieces for easy removal. Practice will enable a person to perform this procedure smoothly without moving the bottle from its position on the tabletop. The label should be facing the host at all times.

2. Remove the foil. Wipe the top of the bottle with an inside corner of the clean service cloth, removing any mold or mildew that might have formed between the cork and the foil capsule (a normal occurrence, especially on older red wines).

3. Using a corkscrew, insert the worm of the corkscrew into the center of the cork. To get the worm centered, the tip must be slightly off center—be sure to center the whole worm and not just the tip.

4. Grasping the corkscrew in the right hand and the neck of the bottle in the left, turn the corkscrew three-quarters of the way into the cork. Turn the corkscrew, not the bottle.

5. Pull the long handle end of the corkscrew down so the lever rests on the lip of the bottle. With one smooth motion, pull the cork straight up and almost out of the bottle.
6. Unscrew the corkscrew.
7. Using the fingers and thumb of the right hand, still grasping the bottle in the left hand, lever the cork from the bottle. Work slowly; avoid dripping and "pops."
8. Smell the cork. If there is any suspect odor, obtain another bottle. Otherwise present the cork to the host.
9. Using another clean corner of the service cloth from the left arm, clean the bottle opening to remove any grime and fragments of cork.

with twisted wire and a foil cap. This protection ensures that the cork stays in the bottle as sparkling wines are under enormous pressure (about 6 atmospheres).

Serving the Wine

Wine is not necessarily served immediately after the bottle is opened. Many quality red wines improve if they are allowed to stand, opened, at dining room temperature for about an hour. As a general rule, champagne and sparkling wines lose some of their quality (the bubbles) if they stand opened, and very old rare wines may oxidize too quickly if opened and allowed to "breathe" for a long period prior to tasting and drinking (see box on pp. 136–137).

Carafes and Decanting Baskets

Restaurants that buy wine in bulk may transfer small amounts to attractive serving pieces for use in the dining room. Usually the wine is poured into a decanter, a carafe, or a specialty bottle with a narrowed neck and a flared mouth called a "liter." The serving procedure is the same as previously outlined once the bottle has been opened.

Decanting baskets or cradles are often used by restaurants for serving red wines. In general, this practice is unnecessary as 90 percent of the red wines being sold and served are ready to drink when released by the producer and do not have sediment requiring a cradle or decanting. However, if it is an aged bottle of red wine, there is a good chance of significant sediment, requiring decanting or special care.

TECHNIQUE FOR OPENING SPARKLING WINES

1. Chill the sparkling wine (in either a cooler or a refrigerator for 24 hours or in a salted ice water bucket for at least 20 minutes) before opening it. Handle the wine carefully—if it is not cold enough or has been shaken, it will "pop" or explode on opening.
2. Carry the bottle carefully, without shaking it, while transporting it to the table. Take the wine from the bucket. Using a clean service cloth, dry the bottle so it will not slip from your grip.
3. Twist the little wire loop that protrudes from the tinfoil until the wire loosens. It should take six twists.
4. Remove the wire. The foil may come off with the wire. If it does not, remove the foil.
5. Once the wire is removed, cover the cork with the service cloth and place the thumb of your right hand on the top of the cork. Keep your thumb on top of the cork at all times—the cork can pop at any time once the wire is removed.
6. Hold the bottle at a 45-degree angle (halfway between straight up and parallel to the floor).
7. Point the bottle away from everybody at the table.
8. Holding the bottle with the left hand, twist the bottle to loosen the cork.
9. As the cork loosens, the pressure will try to force the cork out of the bottle. Put pressure on the cork with your right thumb, keeping the cork in the bottle. Ease the pressure of your right thumb slightly so that *you* control the release of the cork. The cork should slide from the bottle without a pop or geyser, merely a whisper.
10. After the cork is out, keep the bottle at the 45-degree position for 5 seconds to equalize the pressure. Should any bubbles have escaped, catch them with the service cloth, making it appear to the guests that you did a professional job of opening the bottle.
11. Having a cork "pop" should be reserved for those occasions where the guest expressly *wants* the cork to pop (e.g., New Year's Eve). As a rule, and especially with the better champagnes, the cork should whisper and slide out.

PROCEDURE FOR SERVING WINE

1. Approach the host from the right side, holding the bottle in the right hand with a clean folded service cloth draped over the left wrist.

2. Pour a little wine (1-ounce pour) into the host's glass. It will be sipped and either accepted or rejected. Some hosts will wave, meaning that there is no need to taste the wine, that it is fine to pour the guests' wine.

   One of the reasons to pour a taste sip is to determine whether small bits of cork may be floating in the bottle. If there are, the host will be served that wine (and those bits of cork). The guests will be served the "good" wine, without pieces of cork.

   Some guests, usually with European or very traditional upbringing, will expect a tasting glass *and* a separate drinking glass. With this arrangement, even the host does not need to drink wine with cork in it!

3. Proceed to serve the guests, moving in a clockwise direction around the table, if possible. Serve women first, starting with the one on the host's right, and then men, walking clockwise around the table a second time if the male and female guests are alternated.

   It may be the house practice to save time by serving all guests in one pass. In that case, proceed clockwise from the host's left. Serve the host last.

4. Pour the wine from the right side of the guest, using the right hand. Pour carefully and steadily.

5. Hold the bottle so each guest can read the label, with the hand over the back of the bottle, the thumb around one side, and the fingers around the other.

6. Hold the index finger on the shoulder, not the neck, of the bottle.

7. Place the lip of the bottle just over the edge of the glass, but not touching the rim, which is the most fragile part of a glass.

8. Using the wrist, tip the bottle slowly downward until the wine begins to flow.

9. When 3 to 4 ounces of wine have been poured (a standard pour), pivot the wrist to move the hand upward. The glass should be one-third to one-half full for still wines and to within one-half inch of the rim for sparkling wine, if using standard wineglasses (6- to 8-ounce glasses). If using oversize glasses, the fill-level of the wine in each glass

will vary. With practice, you will get accustomed to serving a 3- to 4-ounce pour and will be able to dispense such into any size wineglass.

You can pour six to eight glasses of wine from a standard 750-ml bottle. There are approximately 25 fluid ounces of wine per 750-ml bottle.

Pour champagne or sparkling wines in two movements. The first step is to pour approximately 2 ounces of champagne or sparkling wine into the glass. It will foam, filling the glass. Wait for the foam to subside, then pour additional champagne or sparkling wine to fill the glass within one-half inch of the rim.

10. As the bottle leaves the downward position, give the bottle a gentle quarter twist to avoid dripping wine on the tablecloth.

11. Touch the lip of the bottle to the folded service cloth draped over the left wrist.

12. Replace the wine in the bucket or on the side station or booster station, if one is available. It is acceptable to place red wine on the table. Do not take it out of the room.

13. Refill glasses as they are emptied unless the host has instructed otherwise.

As the bottle nears empty, quietly ask the host if another bottle should be brought to the table. If not, ask the host for instructions on the distribution of the remainder or distribute whatever is left among those guests needing a refill.

Since there may not be sufficient wine to refill all guests' glasses to the proper level without another bottle, it is wise to ask the host for serving instructions—rather than offend someone, it might be easier to order another bottle.

14. When serving a second bottle of wine:

If a *different* kind of wine is ordered, bring clean glasses for everyone at the table.

Bring a new *tasting glass only* for the host if a second (or third) bottle of the same wine is ordered.

When an operation uses decanting baskets or cradles to pour red wines, the server's task becomes much more difficult. First, there is the almost assured risk of shaking sediment in wines that should have been decanted. Second, almost inevitably the wine drips when served from decanting baskets or cradles. The server must have a clean service cloth folded into a pad in the left hand when pouring from a basket. The lip of the bottle must be wiped after every pour and the napkin pad must be held under the bottle while it is being moved from guest to guest. While the baskets can be decorative, most establishments today consider them to be rather bulky, out-of-date accessories that must be inventoried, stored, cleaned, and used only for show. The current trend is either to decant an aged wine with sediment into a decanter or to pour directly from the bottle into the wineglasses if there is no sediment.

❖ LIQUOR LAWS *

General Information

Each state has its own laws relating to the sale of liquor. Each server, bartender, manager, and owner in the hospitality business has an obligation to know and enforce those laws. Each restaurant should prepare a pamphlet or handout to staff members who prepare or serve alcoholic beverages.

This handout should contain information relevant to that establishment and the laws of the state. For instance, in a metropolitan area, it is often prudent to call a taxicab to take an intoxicated patron home or to his or her hotel. In a rural or suburban area where there are few taxis or there will be a half-hour or hour wait for one to arrive, it might be better to try to locate a friend, relative, or neighbor of the person who would be willing to drive the intoxicated patron home.

Items to include in an employee liquor information handout or policy manual are detailed in the chart.

---

*The information in this section is intended as a guide and as training material, so readers may gain a basic understanding of the issues regarding the sale and serving of alcoholic beverages at an establishment as well as some of the legal issues surrounding such service. These materials are provided for informational purposes only and are not to be relied upon. Legal counsel or establishment management, or both, should be consulted to determine how this material relates to each establishment and its unique situations.

SAMPLE LIQUOR INFORMATION HANDOUT
FOR EMPLOYEES

Questions and Answers

Q What is the legal drinking age?

A In many states it is 21 years old.

Q What type of liquor license does the establishment have?
Beer sales only?
Wine and beer sales only?
Wine, beer, and liquor?
To be consumed on the premises only?

A If only on-premise sale and consumption is permitted, all alcoholic beverages sold on the premises must be consumed there. If not consumed, they must be discarded. Neither drinks nor bottles containing alcoholic beverages (opened or unopened) can leave the premises.

   If the establishment has a "package store" or "carry-out" liquor license, the establishment can sell alcoholic beverages to be taken off the premises.

Q What are the laws regarding alcoholic beverage service?

A Persons under the legal drinking age may not be served alcoholic beverages *or* be permitted to consume such beverages on the premises of the licensed operation.

Q How does one determine whether someone is of legal alcoholic beverage drinking age?

A If someone is suspected of being less than 21 years of age, proof of age should be requested.

Q What is "proof of legal drinking age"?

A *Two* picture "ID" cards. Two pieces of identification, at least one with a recent photograph, preferably in color, and with signatures.

   Ask the person for a signature and compare it against the signature on one of the pieces of identification—see if they match.

   Ask the person to validate information on one of the pieces of the picture/photo ID. For instance, ask for the home street address, date of birth, color of eyes, and so on.

Q Suppose parents order a drink for an underage person, saying "He [she] is under our supervision, and we do this at home?"

A This is illegal and cannot be permitted in a public establishment governed by the liquor laws of the state.

   Call a manager to the table. The underage person is not permitted to consume any alcoholic beverage on the restaurant premises regardless of who purchased the beverage and what they do at home.

Q What if the party has not finished the bottle of wine and wishes to take the remainder home with them?

A This is not permitted. The restaurant liquor license permits the sale of alcoholic beverages only for consumption on the premises. Taking any alcoholic beverages off the premises could result in a fine or loss of license.

If the establishment has a carry-out liquor license then a full bottle of wine can be purchased and taken out, but not an open bottle.

Q Suppose a party wishes to take the wine bottle home with them?

A This is permitted as long as the bottle is empty. Take the bottle to the kitchen, dump any remaining wine, beer, or spirits, rinse the bottle with water, and place in a paper bag if available, or wrap in foil. Then return the empty, wrapped bottle to the guest(s).

Q What if a guest becomes and is acting intoxicated?

A If someone becomes intoxicated, or acts as though he or she might be intoxicated, or has consumed a quantity of alcoholic beverages such that an ordinary person of their height and weight would probably be intoxicated, do not serve that guest any further alcoholic beverages.

Call a manager and inform the manager of the situation. The intoxicated guest should be offered a cup of coffee or soda or other nonalcoholic beverage and allowed to remain on the premises, sobering up, if he or she is not unruly.

If the patron is unruly, a taxi should be called and asked to take the patron to his or her hotel or home or place of residence. The establishment should prepay the taxi for the cab ride.

If the patron has a friend, relative, or associate who is able and willing to take the guest home or to a place of his/her choice, management should request the person's phone number. The manager should call and ask the person to pick up the intoxicated person. The intoxicated person should be removed to a quiet area so as not to distract other customers and be provided with nonalcoholic beverages and/or perhaps a complimentary snack while awaiting the ride.

Under no condition should the intoxicated person be permitted to drive a motor vehicle. Someone should be called to pick the person up, or the operation should call a taxi and pay for the person to be taken to a place away from the operation (e.g., home).

SAMPLE LIQUOR HANDOUT
FOR EMPLOYEES

Alcoholic Beverages

"Our Restaurant" [name of restaurant] is proud to serve an extensive offering of wines and other alcoholic beverages to our guests.

Service of alcoholic beverages is governed by both federal and state laws. We adhere to and comply with all laws relating to the sale, service, and consumption of alcoholic beverages.

The legal age for consumption in our state is _____ years of age. If there is any doubt regarding a person's age, ask the person for proof of age. Proof of age can be satisfied by two recent picture/photo ID's. When the photo ID is presented, ask the person questions that are answered by the photo ID such as the person's address, date of birth, or color of eyes. The person's answers should match the data on the ID card. If they don't, don't serve that person alcoholic beverages. Keep the ID card and give it to a manager, explaining what happened. It will be the manager's duty to determine whether the ID is valid for that person and whether or not local police should be called to investigate.

If persons over the age of 21 or legal drinking age order alcoholic beverages and wish to share those beverages with persons under the age of 21 (e.g., their children), it is illegal and cannot be permitted. Notify a manager or maître d' at once. The alcoholic beverage must be removed from the person under the age of 21. What guests do at home is their business, but on the restaurant's premises, no one under the legal alcoholic drinking age may drink alcoholic beverages.

Dram Laws—Responsibility for Intoxicated Persons*

The serving of alcoholic beverages to guests is a privilege that is accorded to them and not a right of each guest.

By law, when a person becomes intoxicated, and the servers of alcoholic beverages to that person know or should know that that person is intoxicated or should be intoxicated (for an average person of that height and weight who has consumed that amount of alcohol in that period of time), the establishment, the bartender, and/or the servers can be held legally liable for consequences resulting from such intoxication.

Signs of intoxication include the following:

◆ Staggering.
◆ Not being able to walk a straight line.
◆ Slurred speech.
◆ Lack of coordination or manual dexterity.
◆ Glazed or bloodshot eyes.
◆ Drowsiness, in conjunction with any of the preceding signs.

The server of the alcohol and/or the bartender, who made the drinks, and/or the establishment can become liable for consequences of the patron's actions resulting from such intoxication because it was aided and abetted by the server's or servers', the bartender's or bartenders', and the establishment's actions.

Therefore, if you suspect any guest of being intoxicated, refuse to serve the person further alcoholic beverages and notify management immediately. It is management's responsibility at that point to handle the situation and to ensure that the guest leaves the premises in such a way as not to harm himself or herself, or others. For example, the manager might call a taxi for the guest so as not to allow the person to drive home.

*This section is intended only as a guide and as training material, so readers may gain a basic understanding of the issues regarding the sale and serving of alcoholic beverages at an establishment and some of the legal issues surrounding such service. These materials are provided for informational purposes only and are not to be relied upon. Legal counsel or establishment management, or both, should be consulted to determine how this material relates to each establishment and its unique situations.

THE "SIZZLE," OR, WHY RETURN

E i g h t

Nonalcoholic Beverages

Is there something unique, different, exciting about your operation that people are willing to travel out of their way for? If so, you have sizzle. If not, it's what you want to get.

How to acquire "sizzle"? There are lots of ways. A hotel might offer child care services for guests, with a supervised play area and organized activities for different age groups ranging from toddlers to teens. Or, a resort might have the "hottest" bands in town, fashion shows in the dining room at various times and on certain days, or cartoon characters walking throughout the property, giving autographs to children.

A restaurant has many options. Part III of this book explores services that guests have been clamoring for and have demonstrated a willingness to pay for (a critical consideration). This chapter discusses nonalcoholic beverages, and later chapters will cover coffee and tea service, extravagant foods, and rare and organic wines.

Nonalcoholic beverage sales are growing. These include fruit-based drinks, nonalcoholic wines, nonalcoholic beers, sodas, and bottled waters, as well as teas and coffees.

People often enjoy the social aspects of gathering and drinking beverages together but, for one reason or another, don't want to imbibe alcohol. Perhaps it's the desire for a healthier lifestyle or the fear of drinking alcohol and then driving that causes guests to choose a nonalcoholic beverage. Other guests may avoid alcohol for medical reasons or because of religious convictions. While people request and drink nonalcoholic beverages for many reasons, the establishment's main concern is simply to have such drinks available for the guests who ask for them.

The sale of nonalcoholic fruit-based drinks, wines, and beers should be a profit center for businesses. Nonalcoholic beverages should be an adjunct area of beverage sales, marketed to switch those guests who order only table water to an alternative beverage: a fruit-based drink, a glass of nonalcoholic wine, a nonalcoholic beer, a soda, or a bottled water. Servers who can persuade a guest to order not a soda but any of the other nonalcoholic beverages often provide increased sales for the establishment through the extra charges for those drinks.

But having sodas is also important. Trendy "new age" soft drinks, including fruit-flavored sodas and "clear sodas," have been predicted to grow to a 10 percent share of the soft-drink market within the next 5 years. Consumers have seen the growth of new "clear" products such as Pepsi's *Crystal Pepsi* and Coke's *Nordic Mist* as well as the naturally flavored *Snapple* drinks and flavored seltzers such as *Original New York Seltzer*, which is offered in such traditional flavors as root beer and cream. It is important not only to keep them available for guests, but to update the available brands regularly to reflect those requested most often.

❖ SPECIALTY DRINKS

Specialty drinks can be the sizzle that brings customers back. If you have the best frozen strawberry daiquiri in town—made from fresh strawberries—served in an oversized glass and garnished with a fresh, baked-on-premises flat cookie, profiterole, or other signature item, and it's priced for value, people will come back.

Rather than offering only an alcoholic specialty drink, it is becoming increasingly popular and profitable to offer a grouping of nonalcoholic specialty drinks. Recipes for a number of fruit-based specialty drinks are provided in this chapter. If there is only one "house specialty drink," make it very special and an excellent perceived value for its price. The house specialty drink must sizzle—guests must want to return for it, to go out of their way for it. The house specialty drink needs to be a marketing tool.

❖ FRUIT-BASED BEVERAGES

Fruit drinks are increasingly popular. The ingredients are readily accessible and the drinks are easy to make with the advent of juicers, juice extractors, blenders, and "slush" machines. The drinks? Your imagination is the limit. Try old favorites such as fresh-squeezed orange juice or lemonade then begin adding new ingredients and various juices. Try new combinations you normally wouldn't think of juicing—fruits and vegetables you would usually boil or bake or eat in raw chunks, such as carrots, celery, and apples. With a juice extractor, any number of fruits and vegetables are at your fingertips as base ingredients. Nutrition-conscious guests will thank you if you also use honey, maple syrup, or artificial sweeteners rather than refined white sugar.

Even without a juice extractor, many wonderful, healthy drinks can be created using a blender. If you want a "frozen daiquiri" type of drink, blend the basic drink with ice. If you want an "ice cream" or "frozen yogurt" drink, combine the drink with ice cream, frozen yogurt, or even fresh yogurt in a blender. For a more traditional cocktail, use a larger quantity of fruit or vegetable juices than of the yogurt or ice cream.

Whatever drinks the house creates, market them. Guests are asking for nutritious foods and drinks; let them know healthy options are available. Use table tents, tableside suggestive selling, menu inserts. There might be a "daily special" or "drink-of-the-day" board; advertise the new, healthy drinks there. Stress the advantages of these wonderful drinks—natural ingredients, low fat, natural fruit sugars, no alcohol.

Market one of these drinks as the "house specialty." Give samples to customers, let them try it. As customers begin to ask for the specialty, add other healthy, nonalcoholic beverages to the ordering options. Once guests begin to ask regularly for the house specialty, put it on your beverage menu and create table tents. After a few weeks, when consumption levels off, establish par levels— the amount of each beverage you need to keep on hand each shift.

Make specialty drinks in large batches to save on the time and effort required. Since fresh fruits and vegetables are used for these drinks, it makes sense to prepare bulk quantities. Many of the recipes that follow are for four gallons. Once you've established your par level (e.g., 4, 8, or 12 gallons), you can plan ahead. Pour the finished drinks into 1-gallon (or smaller) containers and bring to the bar as required. Have the backup easily accessible.

The following recipes include both some newly developed and tested as well as tried-and-true health drinks. Try them as suggested or add some of your own imagination's ingredients. Have a contest for your guests to submit their recipes for a new "house specialty drink."

"Pour and Drink" Drinks*

NONALCOHOLIC KIR   Yield: 1 drink

6 ounces   Nonalcoholic sparkling cider
1 tsp   Raspberry syrup

Pour sparkling cider into glass. Pour raspberry syrup into cider. Garnish with raspberry.

*Note: tsp = teaspoon
Tbsp = Tablespoon

HOT MULLED CIDER   Yield: 1 gallon

(Market it as an alternative to the warm after-dinner brandy tradition. For instance, a server might suggest to guests who have finished their dinner, "Would you care for an after-dinner drink or brandy, or may I suggest our house specialty, 'Swamp Cider,' a nonalcoholic mulled cider served in an oversized brandy snifter?")

1 gallon	Apple cider
8	Cloves, whole, inserted into orange rind
3	Allspice, whole
3 sticks	Cinnamon
1 cup	Brown sugar

Place first four ingredients in large saucepan and bring to a boil. Simmer for 10 minutes. Add brown sugar and simmer another 10 minutes. Keep warm. Serve with cinnamon stick as a garnish.

Variation: Add 1 quart cranberry juice to first four ingredients in saucepan. Proceed as above.

BERRY COCKTAIL   Yield: 2 quarts

1 quart	Strawberries, ripe
25	Oranges

Juice oranges. Wash strawberries; remove greens. Puree strawberries with orange juice. Blend juices together.

MOSTA MIMOSA   Yield: 1 gallon

1 quart	Strawberries
20	Oranges
2 bottles	Sparkling apple cider or nonalcoholic champagne, 750 ml size

Juice oranges. Wash strawberries; reserve one pint for garnish. Remove stems and greens. Puree strawberries in blender* with one cup orange juice (see Figure 8.1). Mix fruit juices. For each cocktail, add equal parts fruit juice with sparkling cider/champagne. Garnish with reserved strawberry.

CITRUS DELIGHT   Yield: 1 gallon

20	Oranges
7	Grapefruit
3	Lemons
3	Limes
2 quarts	Soda water
	Crushed ice (optional)

*Note on blender drinks: We suggest using a hand wand mixer for large batches, such as a "Dynamic Blending Mixer®" (Novon Company, Inc, Pleasanton, CA). Alternatively, prepare drinks in batches if blender has a small capacity.

Juice all citrus. Mix citrus juices together. In a glass, pour in half citrus juices, top with soda water.

Variations: serve over crushed ice or mix in blender with crushed ice. Garnish with mint leaf or lemon/lime wedge.

ZESTY BEST   Yield: 3 quarts

1	Pineapple
25	Oranges
6	Lemons
2 cups	Ice
1/2 pint	Raspberries

Wash raspberries. Puree. Strain. Juice oranges, lemons. Peel core and cube pineapple. In blender, combine pineapple cubes with orange and lemon juices; add ice. Blend until smooth. Drizzle raspberry puree into bottom of glass, add pineapple mixture, top with dot, approximately one-half teaspoon, of raspberry puree.

Variation: Strain blended pineapple, orange, lemon juice. Drizzle raspberry puree into bottom of glass; fill with ice cubes. Pour strained fruit juices over ice.

Prepared juices method:

1 quart	Pineapple juice, fresh
2 quarts	Orange juice, fresh squeezed
1 1/2 cups	Lemon juice, fresh squeezed
2 cups	Ice
1/4 cup	Raspberry puree, strained

Stir together juices. Fill glasses with ice cubes, drizzle raspberry puree into glass, top with fruit juices.

Blender Drinks

FRESH FROZEN BANANA DAIQUIRI   Yield: 6 servings

6	Bananas, ripe
3	Limes
2 Tbsps	Honey
1 quart	Apple cider
2 cups	Ice
tt*	Cinnamon
tt	Nutmeg

Peel bananas, limes. Squeeze and remove seeds from lime juice. In blender, combine bananas, lime juice, honey, apple cider, and ice. Blend. If a "slush" drink is desired, use more ice. If a thinner drink is desired, use additional apple cider. Blend to desired consistency. Flavor with cinnamon, nutmeg.

*tt = to taste.

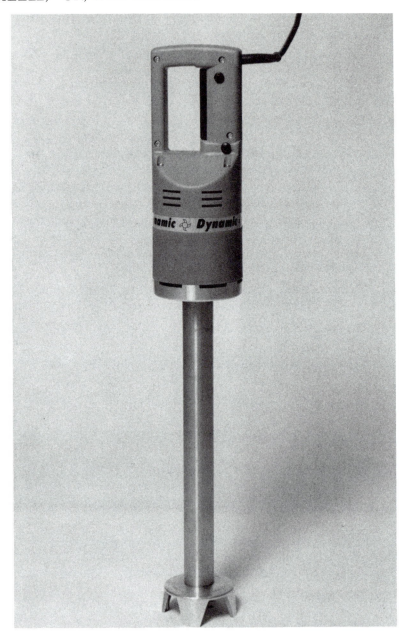

FIGURE 8.1   Dynamic Blending Mixer.®

ZANGY PINES   Yield: 1 gallon

1	Pineapple
5	Granny Smith apples
1 quart	Apple juice or apple cider
1 pint	Ice cubes or crushed ice

Peel, core, and remove eyes from the pineapple. Cut into cubes; reserve several slices for garnish. Peel, seed, and cut apples into cubes. Place pineapple and apple cubes in blender. Add apple juice and ice. Blend. If thinner consistency drink is desired, add more apple juice. If a thicker "slush" drink is desired, use more ice.

FRESH FROZEN STRAWBERRY DAIQUIRI   Yield: 2 quarts

2 pints	Strawberries
1 quart	Apple cider
2 Tbsps	Honey (optional)
2 cups	Ice

Wash strawberries. Reserve 1/2 pint for garnish. Remove greens from remaining strawberries. In blender, combine strawberries, apple cider, and ice. Blend to desired consistency, using more ice to make the drink "slushier" and more apple cider to make it thinner. Use honey to sweeten to taste. Garnish with fresh strawberries.

SOUTH OF THE BORDER TOMATO SPLASH   Yield: 4 gallons

10	Tomatoes, ripe
1 can	Tomato juice, 48-ounce size
1 cup	Cilantro, chopped
3	Lemons
tt	Celery salt
tt	Tabasco
tt	Worcestershire
tt	Hot pepper sauce
1 bunch	Celery

Peel, quarter, and deseed tomatoes. Chop into large chunks. Squeeze the juice from the lemons and remove all seeds. Place tomatoes, cilantro, and lemon juice in the blender and blend until smooth. Add tomato juice. Season to taste. Serve with celery stick.

HONEY, DO   Yield: 1 gallon

1	Honeydew melon, ripe
4	Grapefruit
12	Oranges

Peel, seed, and cube the melon; juice the grapefruits and oranges. In a food processor (or blender if a food processor is not available), puree the honeydew with the grapefruit and orange juices.

Variation: Add ice and blend to desired consistency for a "slush" drink.

STRAWBERRY LIME SLIME

Yield: 2 quarts

1 quart	Strawberries	Syrup:
4	Limes	1 cup water
1$\frac{1}{2}$ cups	Seltzer water	1 cup sugar
$\frac{1}{2}$ cup	Lime syrup	Grated rind from 3 limes
2 cups	Ice	

Make Lime Syrup: Combine water with sugar. Place on high heat, stirring until sugar is dissolved. Add grated lime rind. Cover; bring to a boil for 1 minute; uncover (this dissolves any sugar crystals on the side of the pan). Boil rapidly for 10 minutes total; cool, strain.

Wash strawberries; remove greens. Juice limes. Blend strawberries with lime juice and ice; add lime syrup and seltzer. Garnish with additional strawberries or lime slices.

DOTTED SWISS

Yield: 2 gallons

1	Pineapple
24	Kiwi
1	Honeydew melon
2 quarts	Apple juice
1 pint	Ice

Peel, core, and remove eyes from pineapple. Cut into chunks. Peel kiwi; slice. Peel and deseed melon. Cut into long slices, then chunks. Place in blender with apple juice. Puree until smooth. Add ice, puree again.

Variations: For a thinner consistency drink, use more apple juice; for a "slushier" drink, use more ice.

SATIN SMOOTH

Yield: 2 gallons

1	Pineapple
10	Granny Smith apples
1	Cantaloupe
1 quart	Apple juice
1 pint	Ice

Peel, core, and remove eyes from pineapple. Peel and seed apples and cantaloupe. Slice pineapple and melon into long slices. Place fruits in blender with apple juice and ice. Process until smooth. If thinner consistency drink is desired, use more apple juice; if a thicker consistency drink is desired, use more ice.

Juice Extractor Drinks

SOUTH OF THE BORDER TOMATO SPLASH

Yield: 4 gallons

20	Tomatoes, ripe
1 cup	Cilantro, chopped

3	Lemons
tt	Celery salt
tt	Tabasco
tt	Worcestershire
tt	Hot pepper sauce
1 bunch	Celery

Quarter the tomatoes. Peel the lemons. Juice half of the quartered tomatoes in a juicer. Add cilantro and lemon. Juice remaining tomatoes. Season to taste. Garnish with celery stick.

SUNSET COCKTAIL   Yield: 4 gallons

15	Apples
10	Beets
7	Oranges
21	Carrots, medium

Wash and trim ingredients. Peel carrots. Peel oranges in a spiral design, reserving orange peel. Juice apples, beets, oranges, and carrots together (the order doesn't matter). For garnish, scrape white membrane from inside of orange peel; cut into 4-inch slices. Spiral slices on a long toothpick.

TANGY APPLE COCKTAIL   Yield: 4 gallons

6	Pineapples
24	Granny Smith apples

Peel and core the pineapples. Slice into long slices. Reserve one slice for garnish. Quarter the apples and remove seeds. Juice apples and pineapples. Cut reserved long slice into cubes for garnish.

MELONS AHOY   Yield: 4 gallons

4	Cantaloupes or Crenshaw melons
4	Honeydew melons
1/4	Watermelon
16	Bananas

Peel melons and bananas. Juice together, alternating bananas between the melons. Garnish with slice of melon.

HOT TAMALE   Yield: 4 gallons

30	Green peppers
1 3/4 cups	Cilantro
1/2 cup	Jalapeno peppers
2 ounces	Celery salt
2 cans	Tomato juice, 46-ounce size

Wash peppers, cilantro. Juice cilantro then peppers. Add spices and tomato juice. Garnish with celery sticks, if desired.

THE BEEMER   Yield: 4 gallons

30	Beets
31	Carrots, medium
2¹/₂	Daikons

Wash and peel vegetables. Slice one carrot very thin on a vegetable slicer; place in water, reserving for garnish. Juice remaining vegetables. Curl reserved carrot slices and use as garnish.

Note: If daikons are not available, substitute 2¹/₂ pounds of radishes.

PLANTATION DELIGHT   Yield: 4 gallons

12	Pineapples
2 cups	Honey
4	Lemons
7 pints	Strawberries

Wash strawberries. Reserve one pint for garnishes. Remove stems and leaves from remaining strawberries. Peel, core, and slice pineapples into long slices. Peel lemons. Juice together. Add honey. Mix well. Garnish each drink with a whole strawberry.

GRAPE HEAVEN   Yield: 4 gallons

5 pounds	Green grapes
5 pounds	Red grapes

Wash grapes. Remove stems, wash again. Juice together, alternating green and red grapes. Serve.

Variations: Mix with ginger ale or nonalcoholic sparkling wine; mix with grapefruit juice and serve with ginger ale or sparkling wine; mix in blender with crushed ice (Icy Grape Heaven).

GREEN MACHINE   Yield: 4 gallons

4 pounds	Spinach
10	Carrots, medium
10	Tomatoes, medium, ripe
10	Oranges
1 Tbsp	Garlic
tt	Salt, pepper
tt	Tabasco
tt	Hot sauce

Wash spinach in three changes of cold water. Wash carrots, tomatoes, oranges. Peel carrots, oranges. Juice, alternating spinach between carrots, tomatoes, oranges, garlic. Season to taste.

REVVER UP   Yield: 4 gallons

15	Tomatoes
8	Carrots, medium

3	Beets
1½ cups	Parsley
1 pound	Spinach
4	Lemons
tt	Worcestershire sauce
tt	Tabasco sauce

Wash spinach in three changes of cold water. Chop. Wash vegetables, parsley. Trim and peel carrots and beets. Peel lemons. Juice all ingredients, alternating spinach and parsley between carrots. Season to taste.

BERRY BERRY GOOD   Yield: 4 gallons

5 pounds	Thompson seedless green grapes
10	Carrots, medium
5 pints	Strawberries

Wash grapes, carrots. Remove stems from grapes, rinse with water. Trim and peel carrots. Wash strawberries. Reserve one pint for garnishes. Remove stems and greens from remaining strawberries. Juice.

Serve alone as a cocktail; blended with ice for a "slush" drink; blended with vanilla or strawberry ice cream for an ice cream drink. Garnish with fresh strawberry.

GREEN AYES   Yield: 4 gallons

23	Oranges
4 pounds	Spinach
4 bunches	Parsley
7	Lemons
3	Limes

Wash and slice two limes, two lemons, and two oranges for garnish, sandwiching one quarter-slice of lime between one quarter-slice of orange and one quarter-slice of lemon on a toothpick.

Wash remaining fruits. Wash spinach in three changes of cold water. Steam spinach slightly to wilt. Peel remaining oranges, lemons, and lime. Juice, alternating spinach between oranges.

NIRVANA   Yield: 4 gallons

12	Carrots
7	Oranges
6	Apples
10 stalks	Celery
3	Lemons

Wash items. Trim and peel carrots, oranges, lemons, apples. Remove seeds from apples. Juice all ingredients. Garnish with additional celery sticks.

KIWI SPARKLE   Yield: 4 gallons

6	Pineapples
15	Kiwi
3 pints	Strawberries

Peel and core pineapples; slice into long strips. Reserve three strips for garnish. Peel kiwi. Reserve three kiwi for garnish. Wash strawberries. Reserve one pint for garnish. Remove greens from remaining strawberries. Juice pineapples, remaining kiwi, and strawberries. Garnish with kiwi slice backed by pineapple slice, topped with a strawberry.

ELIXIR   Yield: 1 quart

1 bunch	Celery
3	Granny Smith apples
4	Carrots, medium

Wash all vegetables. Core apples; quarter. Peel carrots. Juice together. Garnish with additional celery stalks.

PINEAPPLE DELIGHT   Yield: 1 gallon

1	Pineapple
12	Kiwi
2 pounds	Grapes, green

Wash grapes and remove stems, wash again. Peel kiwi. Peel and core pineapple, slice into long pieces. Juice, alternating fruits. Garnish with pineapple slice.

CARROT MEDLEY   Yield: 2 quarts

1	Pineapple
1	Honeydew melon
6	Carrots

Wash and peel carrots. Peel honeydew, remove seeds, slice into long pieces. Peel and core pineapple, slice into long pieces. Juice, alternating carrots with the fruits. Garnish with pineapple slices.

"Health Cream" Drinks

Ice cream is loaded with heavy cream, which many people are trying to minimize in their daily food intake. As a substitute, use low-fat, low-sugar frozen dessert substitutes in traditional "ice cream" drink recipes.

The following recipe is for a healthy ice cream substitute.

HEALTH CREAM

Yield: 2¹⁄₂ quarts	Yield: 4 gallons	Ingredients:
5 cups	2 pounds	Low-fat yogurt
2¹⁄₂ cups	2 pounds	Part-skim ricotta cheese
2 cups	1 quart	Honey or maple syrup
¹⁄₄ gallon	¹⁄₂ gallon	Ice cream mix
2 Tbsps	¹⁄₄ cup	Vanilla

Optional: Puree ricotta cheese to make a creamier, smoother mix.

Combine all ingredients; put into ice cream maker and freeze according to manufacturer's directions.

Note: The ice cream mix will stabilize the mixture so it can be used for approximately 1 week.

Fruit Flavors: Add fruit puree to the mixture before placing in ice cream maker.

Cinnamon: Mix together, then on high heat reduce to one-half original volume:

Yield: 2¹⁄₂ quarts	Yield: 4-gallons	Ingredients:
1¹⁄₂ cups	1 quart	Sauterne
2 Tbsps	1 cup	Brandy
4	8	Cinnamon sticks
¹⁄₂	1	Nutmeg, whole
5	10	Cloves, whole
1 Tbsp	¹⁄₄ cup	Brown sugar

After this has been reduced by one-half, cool. Fold into mixture before placing in ice cream maker.

FRUIT SMOOTHIE   Yield: 1 gallon

2	Pineapples
8	Bananas, ripe
20	Oranges
1 quart	Yogurt, plain or Health Cream
¹⁄₄ cup	Honey

Juice pineapples and oranges. In a blender, blend bananas and juices until smooth, add yogurt and honey to taste. Garnish with fresh pineapple chunk, with strawberry atop (strawberry optional).

Prepared juices method:

1¹⁄₂ quarts	Pineapple juice
8	Bananas, ripe
1 quart	Orange juice, fresh squeezed
1 quart	Yogurt, plain or Health Cream
¹⁄₄ cup	Honey

Blend all ingredients until smooth and blended. Garnish glasses with pineapple chunk, with strawberry atop (strawberry optional).

❖   # BOTTLED WATERS

Bottled waters have become a "hot" commodity. Many people are choosing to consume healthy beverages rather than alcohol, especially during lunch and/or business meetings. In addition, many nutritionists recommend that each person drink at least eight glasses of water per day. It's no wonder that bottled waters have become a growing business.

While sodas still are the number one beverage consumed in the United States (an estimated 13 billion gallons annually), waters are quickly gaining in popularity; sales have increased fivefold since the 1970s. (In terms of consumption, sodas are followed by coffee, beer, milk, then waters.) The consumption of bottled waters grew steadily over the past decade and has rocketed during the past few years. Consumption of bottled waters is expected to be at 20 gallons per capita by the year 2000, up from approximately 8 gallons per capita in 1990. At some establishments, bottled waters comprise approximately 25 percent of beverage sales during lunch, unheard of a few years ago.

Many people list the taste of bottled waters, compared with the taste of local tap water, as the main reason they purchase them; others are concerned about health issues and the chemicals, minerals, and other dissolved substances in municipal water. With lead pipes creating a lead exposure problem for tap-water drinkers, many people without filters on their home tap system are drinking bottled water at home and continuing the practice while away from home to ensure that they know the composition of their drinking water. Others enjoy the bit of carbonation or "sparkle" in the sparkling waters, yet others prefer the flavor of their favorite water.

More than 700 bottled waters are available in the United States alone—still waters, sparkling waters, flavored waters, mineral waters. With these choices, consumers have created a three-billion-dollar per year industry. Consumers often ask for their favorite brand by name—if you're planning to have bottled waters on the beverage menu, it is advantageous to list several. The marketing of waters is fiercely competitive; by offering several waters on a beverage menu, your establishment can reap the benefits of the marketing campaigns of the brands offered. With over 400 bottling facilities in the United States, many of the bottled waters are sold locally and not shipped outside a three- or four-state area. However, some are known (and distributed) nationally, and even internationally.

Mineral waters include some of the best-known bottled waters: *Artesia* (the increasingly popular Texas sparkling water), *Perrier*,

Evian, San Pellegrino, Voslau (the "champagne" of bottled waters, from Austria), *Loka,* and *Swiss Altima.* These waters range from being "still" to having small bubbles and moderate effervescence, and from having zero to moderate sodium content and low to moderate mineral content. Most have a crisp finish to the palate, and all can be placed in that nebulous category of "healthy beverages." Added to these are the new "gourmet, fitness-conscious, nutritious" waters, including the juice-added, fruit, and herb-flavored waters. These waters have natural juices of fruits and/or herbs added, or fruit essences and/or herb essences. Many waters have only the natural sugars of fruit juices (fructose) added. Some brands, however, add corn syrup or other nonfruit sugars; it is advisable to check the label for nutritional information regarding the flavoring.

Marketing ideas range from water tastings to table tents. A restaurant could feature a "Water of the Month" and suggestively sell that water to its patrons. A restaurant could host community meetings, offering waters that are on its beverage menu. After a series of wine tastings, a logical follow-on would be a series of water tastings where tasters compare various brands in blind tastings—perhaps one water tasting for flavored waters, one for still waters, and one for effervescent waters. At catered functions or banquets, have a separate water bar set up or have bottled waters available at the alcoholic beverage bar stations.

What are the categories of bottled waters? Generally, there are mineral waters, spring waters, sparkling waters, still waters, and flavored waters, although many waters fit into two categories; for example, *Evian* is both a mineral water and a spring water; *San Pellegrino* is both a mineral water and a sparkling (effervescent) water.

A basic difference between waters is that they are either sparkling or nonsparkling (still) waters. Sparkling waters, such as *San Pellegrino,* generally don't have calories or caffeine, two considerations important to health-conscious individuals, but do have an effervescence. They are primarily marketed to upscale consumers in lieu of other beverages and are sold in small packages (bottles of 1 liter or less). Sparkling waters are refreshment beverages, consumed not necessarily to quench thirst, but as a beverage with a meal or in lieu of an alcoholic drink or fruit juice at social outings.

Sparkling waters are plain as well as flavored. Flavored waters often have a tint reflecting the additive flavoring: Orange-flavored waters may be slightly orange; raspberry- or berry-flavored waters may have a pinkish tinge. Besides the colorings, calories as well as caffeine may have been added.

Sparkling waters are characterized by effervescence, or bubbles. Some have tiny bubbles—Voslau (Austria) is noted for its tiny bubbles. San Pellegrino (Italy), Loka (Sweden), Artesia (Texas), Perrier (France), and Crystal Geyser (California's Napa Valley) are all popular sparkling waters. While many of the sparkling waters are effervescent, they may or may not have carbonation added. Artesia, for instance, notes on its label that it is noncarbonated. Seltzers and club soda are still waters that have carbonation added but are properly considered soft drinks (by the International Bottled Water Association). Seltzers and club soda are not required to follow bottled water standards; they are regulated separately from bottled waters.

Flavored waters are those with flavorings added. Typically, the flavors are fruit extracts, either juices or essences of juices and/or essences of herbs or spices (e.g., mint, cinnamon). Purists regard only those waters with essences and fruit extracts added as flavored waters; others that have sugars (e.g., corn syrup) added may be regarded as flavored drinks. Flavored seltzers can fall into either category, depending on the manufacturing process. Many flavored seltzers are on the market today, including berries (strawberry, raspberry, blackberry), orange, lemon, lemon/lime, lime, peach, cherry, loganberry, tangerine, lime/kiwi, and root beer. The nutritional labeling will indicate whether or not corn syrup has been added to the flavorings.

Some nonsparkling, or still, waters, such as Evian, are also marketed to the "upscale" market. Still waters can contain minerals; they can be pumped well water, artesian well water, or even distilled waters. These waters can be used not only as an "upscale refreshment beverage" but also as a replacement for tap water. As a replacement for tap water, these bottled waters are used for everyday drinking, cooking, water for making coffee or teas, or even mixed with powdered concentrates for flavored beverages. It is estimated by the International Bottled Water Association that over 99 percent of the water used by households is used for nonconsumption purposes such as washing clothing, washing dishes, bathing, flushing toilets, watering the lawn, and washing the car. Only an estimated 1 percent of a household's water supply is used for drinking or consumption (e.g., cooking, reconstituting powdered mixes).

Of the approximately 8 gallons of bottled water consumed per capita in 1990, approximately 7 gallons were nonsparkling and only one gallon was sparkling water. Nonsparkling waters are by far the leading waters in sales. These include waters taken from municipal water supplies and bottled; waters taken from springs, artesian wells, or pumped water wells; water that has been distilled; and other purified waters. Only about 25 percent of the bottled waters are from municipal water supplies; most (approximately 75 percent) are from natural springs or wells. Bottled waters taken from municipal water supplies must meet the requirements of the *Safe Drinking Water Act.*

Most of the nonsparkling bottled waters are sold in bottles of one gallon or larger. Much of it is distributed through delivery systems that truck it and then deliver it to homes or offices, often on a weekly basis; or consumers purchase the water at grocery stores. Most sales are not derived from restaurants or other eateries.

There are natural waters and processed, purified waters. The natural waters include spring waters, mineral waters, and artesian, or other well waters. If it is from a natural source and not from a municipal or other public water supply, it is considered a natural water as long as it has not been modified, blended, or otherwise altered from its original state. No additions or deletions of substances, including minerals or other dissolved substances, are permitted except for ozonation or other disinfection and/or filtration processes necessary for the safety of those drinking the waters. Many of the natural water sources are protected sources. The bottling company will purchase surrounding land and physically protect its source from contaminants and people who might wish to use the source for personal purposes such as swimming, boating, or fishing. Companies will also attempt to protect the source from geographic contamination, for instance, by preventing others from drilling wells that reach into their water source or disposing of contaminants that might seep into the water supply from a nearby location.

"Spring water" is a subcategory of natural bottled waters. Spring water has been taken from underground springs or underground formations from which water flows naturally to the surface. With no additional processing (other than the bottling and labeling process), they are often touted as "natural spring waters." Evian is the leader in this category, fitting three categories: it is a still mineral spring water. *Spa* (Belgium), *Volvic* (France), and *Naya* (Canada) are other popular natural spring waters.

"Well water" is another subcategory of natural waters. It is taken from wells—holes that are drilled, bored, or dug into the ground—and then is bottled and labeled.

"Drinking water" is a more "wide-open" category of bottled water. Basically it is water, most often originating from municipal water supplies, that has undergone filtration treatments such as passing through activated carbon. The water is then disinfected and bottled. Ozone, a form of oxygen, is the final disinfectant used by most bottling companies. Ozone will disinfect water without leaving any chemical residue, aftertaste, or smell. Prepared drinking water is usually bottled in gallon containers. It is generally considered a processed water.

"Purified waters" are waters that have undergone processes that meet the definition of purified water according to certain standards listed in the *United States Pharmacopeia* and meeting the standards of the *Environmental Protection Agency* (EPA). The water may have undergone reverse osmosis, distillation, deionization, or other processes designed to make the water safe, drinkable, free of odors, colors, and dissolved particles within the EPA and *Pharmacopeia* guidelines. As mentioned, most purified waters are disinfected with ozone to ensure there is no chemical residue, aftertaste, or smell, as opposed to municipalities and other public water supplies that use chlorine as the disinfectant.

Reverse osmosis is a process by which water is forced, under pressure, through membranes or filters that remove 90 percent or more of any dissolved minerals and other dissolved solids. Distillation is the process whereby the water is vaporized, by being boiled until it turns into steam. The steam rises, is collected, then condensed. The condensed water is free of dissolved minerals and other solids because they did not vaporize with the water. Deionization is the process whereby water is passed through resins that remove the majority of the dissolved minerals.

Many people, however, desire certain minerals in their drinking water and do not want the purified waters that are available today. Many natural waters contain dissolved minerals and are thus natural mineral waters. Several marketing campaigns attribute good health to the drinking of those mineral waters, believing, or wishing potential buyers of their product to believe, that the minerals will help a person stay healthy and fit. Health claims for bottled waters must be substantiated to be used in advertising, which explains why many health benefits are alluded to in advertisements and not claimed outright.

Interestingly, the *Federal Food and Drug Administration* (FDA) has regulated bottled waters as a "food." Federal regulations concerning bottled waters, in addition to the Safe Drinking Water Act, include the FDA's *Good Manufacturing Practices* and the *Federal Food, Drug, and Cosmetic Act* health and safety standards as well as the labeling requirements. In addition to meeting federal standards, several states and foreign governments have regulated bottled water content and processing. In particular, California (approximately 35% of the U.S. bottled water consumption), New York (approximately 7% of the U.S. bottled water consumption), and Florida (approximately 6% of the U.S. bottled water consumption) regulate the manufacture, bottling, and distribution of bottled waters.

"Mineral waters" are those waters, natural or processed, that contain dissolved mineral solids. The measure used by regulatory agencies is milligrams of "totally dissolved mineral solids" or TDS (totally dissolved solids) per liter. A mineral water considered to have relatively low totally dissolved mineral solids is in the range of 500 TDS per liter. This 500 TDS per liter, or 500 parts per million (ppm) of totally dissolved solids is a minimum figure that several states have enacted into regulations defining a mineral water. A water with a high mineral content will be in the range of 1,500 TDS per liter. Minerals dissolved in the waters include sodium, calcium, magnesium, potassium, and small percentages of others, including benzine, a mineral that caused the Perrier recall in 1990. Mineral waters include San Pellegrino, Evian, Voslau, Loka, Crystal Geyser, Swiss Altima.

EPA guidelines list, in the *Primary Metals Drinking Standards* and *Secondary Metals Drinking Standards*, metals that should not, to any great degree, be found in drinking water. The specific metals and the allowable levels for each metal are defined by the EPA. The *primary* metals are the more toxic, with very low allowable levels— metals such as antimony, arsenic, barium, cadmium, copper, lead, mercury, nickel, and inorganics such as fluorides and nitrates. The *secondary* metals are those that are less toxic than primary metals. Secondary metals include aluminum, iron, silver, zinc, and manganese.

Mineral waters must have the designation "mineral water" on their labels. Further, the total dissolved solids (TDS) per liter of natural mineral water must appear on the label and be stated in milligrams per liter. Mineral waters are, by definition (as defined by the International Bottled Water Association), natural waters. As such, the bottler must be able to identify and clearly distinguish the specific mineral and trace element content in the water at its point of emergence from the ground, and this composition of the water must be a distinguishing and constant feature of that bottler's water.

Some bottled mineral waters are still and some are sparkling. Of the sparkling, there can be naturally effervescent mineral waters, which can be labeled "naturally carbonated mineral water" or "naturally sparkling mineral water," or those with artificial carbonation added. If the carbonation is added, the label might read, "sparkling natural mineral water," or "carbonated natural mineral water." The similarity of the labels can be quite confusing. Note the subtle difference in "naturally sparkling mineral water" versus "sparkling natural mineral water." With but the change of the first two words, the whole meaning has changed. The first is a naturally effervescent mineral water; the second has the carbonation added to a natural still mineral water. There is yet a third category, "carbonated water" or "sparkling water," which has carbonation added to water, but not necessarily natural water. This might be carbonated distilled or purified water, or carbonated water obtained from a municipal or public source.

The chart lists bottled waters commonly found in the United States.

SAMPLING OF BOTTLED WATERS		
Name	*Origin*	*Type of Water*
Appollinaris	Germany	Sparkling mineral water
Artesia	Texas	Sparkling mineral water
Clearly Canadian	Canada	Sparkling water; flavors
Crystal Bay	USA	Sparkling water; flavors
Evian	France	Still mineral water
Gerolsteiner	Germany	Still mineral water
Ice Mountain	Connecticut	Sparkling spring water; flavors
Loka	Sweden	Sparkling mineral water
Mendocino	California	Sparkling mineral water
Mendota Springs	Minnesota	Sparkling mineral water; flavors
Mistic	New York	Sparkling water; flavors
Mountain Valley	USA	Still spring water
Naya	Canada	Still spring water
Perrier	France	Sparkling mineral water; flavors
Poland Spring	Maine	Sparkling spring water; flavors
Quest	California	Sparkling spring water; flavors
Quibell	West Virginia	Sparkling mineral water
San Pellegrino	Italy	Sparkling mineral water
Saratoga	New York	Sparkling spring water; flavors
Spa	Belgium	
Surgiva	Italy	Sparkling & still spring water
Swiss Altima	Switzerland	Pear-flavored mineral
Vermont Rose	Vermont	Still natural spring water
Vittel	France	Still spring water
Volvic	France	Still spring water
Voslau	Austria	Sparkling mineral water

❖ NONALCOHOLIC WINES

Wines typically contain between 11 percent and 14 percent alcohol; some Beaujolais, some Riesling Kabinetts, and other wines are as low as 7 percent alcohol. With more and more people desiring nonalcoholic beverages, the wine industry has developed a process for creating dealcoholized wines (wine with over 99 percent of the alcohol removed).

A recent study stated that slightly over half of all American adults consume alcohol, and half of those surveyed said that a goal of theirs was to reduce their alcohol consumption in the next year. Because of health issues, social consciousness, and stricter "drinking-and-driving" rules, fewer and fewer people are drinking alcoholic beverages. Those who are still drinking alcoholic beverages are drinking less often and, with rare exception, drinking lesser amounts during each occasion. A National Restaurant Association survey conducted in 1987 found that dealcoholized wines accounted for up to 17 percent of total wine consumption in restaurants where it was available. The message seems clear: If dealcoholized wines are available, they will sell.

Nonalcoholic wines have been growing in popularity among guests at social functions who want to drink something other than fruit drinks or sodas or mineral waters, but don't want alcohol. Sophisticated drinkers may choose the nonalcoholic wines for health and fitness reasons—many of these wines contain approximately half or less than half the calories of regular wine. Pregnant women, people with diabetes, and guests who are on medication may choose nonalcoholic wines as an alternative to other beverages. And even wine connoisseurs don't drink wine 24 hours a day. Many enjoy a break from drinking wines with alcohol but yet desire and enjoy drinking a beverage that still looks and tastes like wine.

The goal of creating a nonalcoholic wine is simple, but the process complex. Ideally, the winegrower wants to use fine grapes and traditional winemaking processes and, while being gentle to the wine, remove the alcohol without significantly altering the wine's flavor, bouquet, or character. Quite a goal. In the past, several attempts were made that did not come close to success. (It is reputed that Carl Jung was one of the first to create a nonalcoholic wine, in 1912.) Today, there are winemakers who are approaching this goal of creating finely made, excellent tasting dealcoholized wines, if they have not already reached it.

According to the U.S. Department of Treasury's Bureau of Alcohol, Tobacco and Firearms (ATF), dealcoholized wine (or nonalcoholic wine, as it is sometimes referred to) may not contain more than 0.5 percent alcohol, a level that occurs naturally in fresh orange juice, many other fruit juices, and some yeast-based products, such as certain breads. Nonalcoholic wines are not truly alcohol-free; they contain trace amounts of alcohol.

One of the first pioneers in this process of dealcoholizing premium wines was *Ariel*, a California winemaker. The wines are made using traditional winemaking methods and using traditional wine varieties, (e.g., Chardonnay, Riesling, Chenin Blanc, Cabernet Sauvignon grape). Traditional harvesting methods, crushing, barreling, and fermenting processes are utilized. After the aging process, the alcohol is filtered out of the wine. Ariel and many other premium winemakers use a cold filtration process to retain as much flavor as possible of the original wine while they are removing the alcohol. Ariel prides itself on making sophisticated-tasting wines using premium grape varieties. There are other winemakers who make dealcoholized wines using the cold filtration method with table wine grapes. These wines, while not having the sophisticated taste of a premium dealcoholized wine, are nevertheless quality table wines and appeal to those who desire a nonalcoholic wine without paying a premium price. Prices of premium dealcoholized wines are about the same as the prices of similar wines with alcohol, which is actually a bargain since dealcoholized wines start out as regular wine and extra processing is required to take out the alcohol!

An older process to dealcoholize wine, distillation, has been used by winemakers, but with much less success. The distillation process removes the alcohol with heat, which also removes many of the wine's flavors and aromas.

Many wine aficionados will argue that the alcohol itself is a major taste component of fine wines and a wine without a minimum of 7 percent alcohol is not a wine and does not possess the "character of the grape," as defined in an alcoholized wine. Some people have called this "mouth-feel," a sense that the alcohol imparts to the mouth as a part of the tasting of the wine—the wine's "backbone" or "depth" or "body."

Nevertheless, several Ariel dealcoholized wines have won gold, silver, and bronze medals when judged against regular, alcoholized wines. One of the first such instances was in 1986 at the Los Angeles County Fair when Ariel entered its Ariel Blanc dealcoholized wine into a miscellaneous white wine competition category. It won a gold medal. Again in 1989, the Ariel Blanc wine won two gold medals in its category against other white wines with alcohol. Ariel's dealcoholized wines have won medals in winetastings against wines with and without alcohol many times since. Modern quality methods of producing a dealcoholized wine are satisfactory to many wine connoisseurs and judges for general wine-drinking purposes.

Types of wine made without alcohol include red and white table wines as well as white Zinfandel, Cabernet, Sauvignon, Chardonnay, Riesling, and blends of these grape varieties as well as "champagnes" or sparkling wines. Virtually every style of wine, every grape variety, is available, or can be made available, in a dealcoholized wine. The technology is available, in use, and constantly being improved.

Ariel's filtration process, sometimes referred to as "ultrafiltration," is a form of reverse osmosis. This method is also used to purify water by removing dissolved solids so the water can be sold as bottled water. To achieve ultrafiltration, the wine is forced to flow under high pressure along a porous cylindrical wall of one or two membranes with tiny pores. Water and alcohol, having smaller molecules than the remainder of the wine juice, or must, pass through this membrane wall leaving a very thick, syrupy concentrate of the wine's flavors and essences in the cylinder. Water is added back to the concentrated wine flavors and essences, creating an almost-reconstituted wine. It lacks only the alcohol.

In addition to Ariel, St. Regis, Carl Jung, Giovane, Sante, Firestone Winery, Paul Masson, and Eisberg wineries produce dealcoholized wines. "Low-alcohol" wines are also available, which contain less than 7 percent alcohol. Some of these beverages are blends of slightly fermented wine added to grape juice and water, similar to spritzers. Many "low-alcohol" wines are marketed as "wine coolers" and bear little resemblance to true wine, being closer to spritzers than to wines made in the traditional manner from traditional wine grapes.

Marketing nonalcoholic wines is usually a complement, or "add-on," to the alcoholic beverage business. Restaurants often market the nonalcoholic wines for social occasions. Perhaps there is a wedding, bar mitzvah, or other significant social event being planned where some of the guests who will be attending the event don't drink alcohol or where a family, for religious reasons, doesn't want alcoholic beverages served. The person contracting for this event, however, wants to do a toast. The banquet manager could suggest sparkling fruit juices or, for those who desire a "champagne toast," a nonalcoholic sparkling wine either in addition to or in lieu of the traditional champagne toast.

Another area of sales potential is to those who have been drinking or are expected to be drinking alcoholic beverages for an extended period, for instance, during a "Super Bowl" game or playoffs. Servers can be trained to suggest a nonalcoholic wine or beer for every other round, or certainly after the server begins to suspect that one or more of the patrons might be becoming intoxicated.

In such cases, the availability of nonalcoholic beverages might help prevent a lawsuit against the server, the bartender, and/or the establishment. Or, if a lawsuit is filed, the availability of nonalcoholic beverages might become part of the defense. The establishment might want to show that (1) it offers nonalcoholic beverages for sale; (2) in the case in point, the server suggested them to the patrons in lieu of alcoholic beverages; (3) no further alcoholic beverages were served to the table after the point of offering the nonalcoholic beverages. This might help to mitigate the case. (Should any legal dispute or question arise regarding alcoholic beverages and/or their service, consult legal counsel. The preceding is intended only to provide a guide to marketing dealcoholized wines and beverages.)

However the nonalcoholic beverages are marketed, servers must learn how these products taste, how to sell them, and how to upsell them. Recent studies have indicated that pregnant women are one of the growing consumer groups of nonalcoholic wines, while men are the number one consumer of nonalcoholic beers.

❖ NONALCOHOLIC BEERS

Many nonalcoholic beers have had the alcohol removed by reverse osmosis just as with the nonalcoholic wines. However, while many of the dealcoholized wines are pumped, under pressure, through a long plastic tube with two membrane layers (one dense but thin, the other thick but more porous), forcing out water and alcohol, in the making of beer only one layer of plastic film or membrane is generally used.

The reasons people are switching from alcoholic beer to nonalcoholic beer are similar to the reasons people are switching from alcoholic wines to nonalcoholic wines. The growth in nonalcoholic beer consumption, especially to 18–25 year old men, has been attributed in part to stricter drinking-and-driving laws and penalties, as well as higher insurance premiums for men in that age category, especially if there has been an alcohol-involved accident (DUI, DWI, or other charge) against the person's driving license.

The marketing of dealcoholized beer is to fill a niche created by guests. There is a demand for nonalcoholic beverages and it is incumbent on establishments to update their beverage lists to accommodate their customers' beverage requests. If one establishment doesn't, another will. It is another point of difference, or going that one step beyond the basics.

Anheuser-Busch and *Miller* were among the first American manufacturers to create and mass-market a nonalcoholic beer. Anheuser-Busch created *O'Doul's* and Miller created *Sharp's.* There are now several imported nonalcoholic beers as well, including *Buckler, Clausthale, Kaliber,* and *Moussy.*

Ceremonial Coffee and Tea Service

Coffee is an almost universal American finish to a meal although occasionally diners prefer hot tea. Good, freshly brewed coffee as well as maintaining a varied selection of teas, can become signature items at your restaurant.

The elements of providing good coffee are simple: It must be freshly brewed within a half-hour of being served and must be prepared and served in clean equipment and china without residues, stains, or other markings. How to make good coffee is detailed in *Food and Beverage Service.*[*] In this chapter, we will consider how coffee service can be a difference maker in your establishment.

Good coffee service begins with having the necessary accompaniments (sugar, sugar substitute, cream, cream substitute, and on request, milk or low-fat milk) on the table prior to pouring or bringing the coffee to the table, or delivering these items at the same time. As points of difference, some establishments offer individual "pour-paks" of flavored coffee creams (amaretto, mocha, Irish cream, hazelnut).

❖ COFFEE SERVICE

Ideally, coffee and tea should be served in warmed cups. If the operation has sufficient space near the dining room(s), cup warmers should be installed. A cup (or plate) warmer is a large, insulated box that holds china at a warm temperature, generally around 170°F to 185°F. Servers should follow a standard procedure for providing coffee service.

*B. H. Axler and C. A. Litrides, *Food and Beverage Service,* (New York: Wiley, 1990), p. 31.

PROCEDURE FOR SERVING COFFEE

1. Gather cold items first (saucers, sugars, milk, and/or cream); then get the hot coffee and warmed cups.
2. Coffee saucers are procured from the kitchen or service stand, then placed on the beverage tray.
3. Accompaniments (sugar, sugar substitute, cream, and/or milks) are placed on the tray. Spoons, if not already placed on the table for each person who has ordered coffee (or tea) are also placed on the beverage tray.
4. Hot coffee should be poured from the large urn in the kitchen into a silverplated coffee pot or whatever service piece is used to bring the coffee into the dining room. Restaurants often use insulated thermal containers that are appropriate for dining room service.

   If coffee is brewed directly in individual service-pots, usually holding 8 to 12 cups of coffee, this step is unnecessary.
5. The coffee pot is placed on a beverage tray. The beverage tray is carried in the left hand.
6. Coffee (and tea) cups are procured from a cup warmer if one is in use. Otherwise, cups should be procured from the kitchen and warmed with steaming hot water. The easiest way to do this is to use the hot-water spigot of the coffee urn. Generally, this is the middle spigot on a three-spigot machine. Fill the cups with hot, steaming water and allow them to sit for a moment. Then pour out the hot water, wipe each cup dry with a clean towel, stack the cups on the beverage tray, and immediately proceed to the dining room while the cups are still warm and the coffee still hot.
7. Where there is team service and two or more servers are serving the diners in a station, one server places the cups and saucers, spoons, and accompaniments while the other trails immediately behind, filling the coffee cups. If there is not team service, one person must do all the steps.
8. Approach the right side of a woman at the table who requested coffee. Place a warmed cup on a saucer and place both, in one motion, at her right side using your right hand. Check for a coffee spoon. If one is not there, place one at the right side of the guest.

   If a spoon is needed, place the spoon on the table linen and *not* on the saucer unless the establishment is considered "fast food" or specializes in "quick service."

Place accompaniments in the center of the table yet close enough to the guest that she can reach them for her coffee (tea) should she desire to use any of them.

If using a team approach, the server quickly moves around the table in a clockwise direction to the next woman who ordered coffee. The second server, following the first with the pot of freshly brewed coffee, pours the first woman's coffee into her cup.

A server working alone would pour the coffee into the cup while it is still on the beverage tray. The filled coffee cup and saucer would then be placed, in one motion, at the guest's right side. The cup is filled prior to being placed on the saucer so that if any coffee is spilled, it spills onto the tray and not onto the saucer. The guest is assured of a clean, dry saucer.

9. This process is continued until all women have been served, then men are served coffee.

10. If both coffee and tea are ordered by guests at the same table, and if tea is brewed in and served from large teapots and not individual pots, then both coffee and tea can be served simultaneously.

11. If individual pots of hot water are brought to each person who ordered tea, those would be placed on the beverage tray in addition to the other items.

When reaching the first woman who ordered tea, first the empty but warmed teacup, on its saucer, would be placed on the table to her right side. A spoon would be placed on the table if one was not there already. Then the pot of tea or hot water, or both, each on an underliner plate, would be placed to the cup's right. There may be a third plate, usually a B&B plate or a monkey dish, that contains the tea bag if tea bags are used.

Note that with tea service, milk is served, not cream. The question to ask is, "Lemon or milk?" (Individualized tea service is reviewed later in this chapter.)

12. Some operations fill all cups in the kitchen or at a side-stand and then bring them to the table. In this case, first gather the accessories, spoons, and the saucers. Place them on the beverage tray. Warm the cups in the kitchen, then fill them with hot coffee or tea. Approach the table and with the right hand, place the first filled cup on a saucer and serve them in one motion to the first woman, generally the eldest lady, at the table. Place a spoon if necessary, then place the accompaniments within reach. Proceed, walking clockwise around the table if possible, to the next woman. After they all have been served, serve the men.

COMMON COFFEES AND
❖ # COFFEE DRINKS

Being an almost universal after-meal beverage, coffees create an excellent opportunity for a house specialty drink as well as an opportunity for fantastic word-of-mouth advertising when it's known that you offer several varieties of home-roasted, freshly ground, and freshly brewed coffee. (Yes, it is possible to roast your own beans. Roast the beans to various degrees of darkness, the darker beans being the more "full-bodied" and "richer" blends. Offer several varieties to your guests, from a "regular" brew to a "strong, rich, full-bodied" brew.)

A restaurant may limit varieties of coffee to a "Coffee Flavor of the Day [or Week]," or it may keep several coffees on the menu permanently. Some of the more popular flavored coffees include Irish cream, amaretto, cinnamon, mocha, vanilla-nut, and hazelnut.

In addition to offering flavored coffees, various types of roasted coffee beans from around the world could be marketed to guests. Each type has its own unique flavor and can be marketed into a specialty item. Some common coffee types include Kenyan, Colombian, Java, Mocha Java, Tanzanian, Hawaiian Kona, and Blue Mountain.

Various methods of preparation add to the possibilities for special offerings. The most common method of preparation is in an urn, where coffee is passed through a filter, or dripped, and in some cases, percolated. Specialty coffees, such as espresso, can be steamed or, in the case of Turkish coffee, boiled. Coffees can be combined with milk or cream, or with liquors, or with sugar. There are endless variations.

Some of the more popular coffee drinks are described in the following sections. Many of these can be brewed either with caffeinated or with decaffeinated coffee beans that have first been roasted to the requisite degree and then ground to the fineness required for the particular drink being prepared. The following coffee drinks can be ordered with dessert, in lieu of dessert, or after dessert. They can become a signature item or specialty drink of the house.

Espresso Drinks

Espresso is made from very dark-roasted coffee beans. The brew should be very dark and rich, with pungent, deep flavor. Ideally, it is served in a demitasse, or specially designed small coffee cup, accompanied by a demitasse spoon.

To make the coffee, the beans are ground fine and the steam of boiling water is forced through the grounds. In other words, it is brewed under pressure, or "steamed" rather than percolated or water-dripped. A cup of espresso should contain about two fluid ounces, a rather small amount. The espresso brew should be concentrated, and full-bodied, with a caramel-colored froth on top. It is generally served in cups reserved only for espresso with a lemon twist as a garnish. A *doppio espresso* is a double espresso, served in a cappuccino cup.

Cappuccino

Cappuccino is espresso that has been topped with steamed and foamed milk. Traditionally, the espresso is made and poured into a cappuccino cup, which is larger than an espresso cup. The milk is steamed and frothed (regular cold milk must be used to obtain the best froth). The steamed milk is poured atop the espresso; then the froth is carefully spooned atop the espresso and milk mixture. An alternative method is to foam the milk first, pour it into the cup, and then spoon the froth atop the steamed milk. Next, make the espresso and, very carefully, pour it in a thin but constant stream through the center of the steamed milk and foam. Garnish with the guest's choice of chocolate shavings, powdered fresh nutmeg, or powdered cinnamon.

Café au Lait, Caffè Latte, Café con Leche

In a cappuccino or other large cup, mix espresso with an equal amount of steamed or scalded milk. This mixture, served with fresh croissants, is a particular favorite with the French in the morning.

Caffè Marocchino

This is an espresso served with some milk, but less than in a traditional caffè latte. It is therefore darker than a caffè latte and is served in a slightly smaller cup. However, it contains more milk than a *caffè macchiato.*

Espresso Macchiato

This espresso is prepared with just a bit or "splash" of foamed milk. It is also referred to as espresso "stained" with a little steamed milk. It is served in a demitasse.

Latte Macchiato

Since this is steamed milk with just a bit or "splash" of espresso, it is an excellent alternative to suggest when there are children at a table with adults and the adults are ordering espresso-based or coffee-based drinks. The children may want the same drink the adults are ordering, but the adults don't want the children to have the caffeine in the espresso or coffee-based drinks.

Espresso Ristretto

A very concentrated coffee, espresso ristretto is made by using less water than in a regular espresso. A *doppio ristretto* is a double ristretto.

Caffè Americano or Caffè Lungo

This is a weak espresso, brewed with more water than a regular espresso or brewed and then slightly diluted. It is served in a demitasse cup.

Caffè Moccachino

Chocolate cappuccino is made by placing a shot of chocolate syrup in a cappuccino cup and then finishing as if making cappuccino.

Caffè Corretto

The espresso is "corrected" with a bit of liquor. Traditional liquors used include ouzo, sambuca, brandy, grappa, or rum. Sometimes the grappa, ouzo, or sambuca is served on the side to be poured into the espresso by the guest.

Caffè con Panna

The espresso is served with whipped cream.

Caffè Freddo

"Cold espresso" is espresso chilled with cold water in a glass.

Turkish Coffee

Turkish coffee is very, very strong. It is traditionally made by boiling finely ground coffee beans (Turkish coffee grind) in a particular manner in a particular pot, called a "briki" in Greek. This pot is

---

PROCEDURE FOR PREPARING TURKISH COFFEE

1. Fill a demitasse with water. Pour one cup of water for every cup of coffee to be made into the brass Turkish coffee pot, (the briki).
2. Measure one heaping teaspoon of Turkish-ground coffee for each cup of coffee to be made and place the coffee directly into the briki.
3. Measure one heaping teaspoon of sugar for every cup of coffee to be made and place the sugar into the pot.
4. Place the briki on the flame and bring the mixture to a boil. Once it has reached a boil, remove it from the flame and allow the bubbling to subside.

   Once the boiling has subsided, place the briki back on the flame and bring to a boil again.

   Repeat. The coffee is brought to a boil three times and removed from the flame three times.
5. Once the coffee has boiled three times, pour it into each cup. Serve with lemon peel as a garnish.

---

usually small and made of brass with a long handle so the hair on the user's forearm isn't scorched in the process of brewing coffee over an open flame. The pot has a spout to pour the finished coffee into the cup. Brikis are made in various sizes for preparing one cup of coffee, two cups, or more.

Turkish coffee will be very thick, and the coffee grounds will sink to the bottom of the cup. The coffee will first taste like a strong, slightly sweet coffee, then a syrupy coffee. When the coffee becomes gritty, the grounds have been reached and no further coffee should be consumed.

When the liquid portion of the Turkish coffee has been drunk, Turkish custom is to turn the cup upside down, quickly, onto the saucer. Allow the coffee sediment to drip from the cup for a minute or two, then turn the cup over again. The remaining configuration of the coffee grounds inside the cup can be "read," telling the drinker's fortune.

Alcoholic Coffee Drinks

Basic Liquor-Coffee Combinations

Most coffee and liquor drinks are prepared in a festive manner. Perhaps they are served in a tall, "Irish coffee" mug, or an oversized wineglass. Usually, they are finished with whipped cream, which should be freshly prepared and not squirted from a can. Sometimes a bit of the liquor used in the drink is drizzled on top of the whipped cream.

To prepare the coffee and liquor drink, place the liquor in the glass first, then add a teaspoon of sugar (optional), then finish with hot coffee. If the drinks are garnished with whipped cream, spoon it on top and add a bit of the liquor atop the whipped cream (optional). Some of the liquors often drizzled on top include green crème de menthe, Kahlua, and Tia Maria.

Popular liquor and coffee drinks include the following:

♦ Irish Coffee, made with Jamison's Irish Whiskey
♦ Mexican Coffee, made with Kahlua
♦ Jamaican Coffee, made with Tia Maria
♦ Italian Coffee, made with Amaretto, Sambuca, and Bailey's Irish Cream
♦ Kioki Coffee, made with Cognac and Kahlua
♦ Nutty Coffee, made with Frangelico and Kahlua

Coffee drinks should be served on an underliner plate and doily, served with a spoon as well as a stirrer. Often, the whipped cream drips down the side of the glass and the underliner and doily will both absorb and catch any drips. The underliner also provides a place for the guest to place the spoon and/or stirrer after stirring the coffee.

Flaming Coffee Drinks

Flaming coffee beverages are a wonderful finish to any meal. They require showmanship and daring as well as care so no one gets hurt as the procedure is dangerous.

Special equipment is required, namely tempered glassware so the glasses can withstand the heat of extremely hot coffee or even the direct heat of the burner, and a gueridon (serving cart for the dining room) with a réchaud or heat source.

FLAMING COFFEE DRINKS USING TWO PANS   Yield: 2 servings

Equipment

Réchaud or two-burner heat source
Two chafing dishes
One ladle
Dramatic glassware for the finished coffee drinks
Gueridon

Ingredients

2 cups	Hot coffee
1/3 cup	Brown sugar
1 lump	White sugar
	Peel from half an orange, in one long strip
	Peel from half a lemon, in one long strip
2 or 3	Whole cloves
1 stick	Cinnamon
4 ounces	Cognac or other quality brandy

Procedure

Wheel the gueridon next to the guests' table. The gueridon should have all ingredients and equipment on it (there are usually two shelves; many of the ingredients and pieces of equipment can be on the lower shelf with the réchaud or heating element on the top).

Light both burners and place the chafing dishes atop the burners. Pour the coffee into one chafing dish to keep it hot. (If there is only one burner, bring the coffee to the gueridon in a thermal container so it retains its heat.)

Place the cognac, brown sugar, lemon and orange peel, cloves, and cinnamon stick in the other chafing dish and heat.

When the brandy is warm, dip in the ladle and remove about 1/2 ounce (about 1 tablespoon of liquid).

Place the sugar cube in the ladle and ignite it with the open flame from the burner, being careful not to burn anyone or anything.

Lower the flaming ladle into the chafing dish of brandy, which will ignite. While it is still burning, pour it slowly and carefully into the chafing dish of coffee from a height of about 24 inches (or from whatever height is comfortable).

Note: This is best performed where the lights are dim, so the flaming cognac is visible. If the lighting in the room is fairly bright, the effect of the flame will be lost. In this case, you can simply pour the coffee into the chafing dish of flaming cognac. It will taste exactly the same, and if the guests can't see the flame very well, it makes no sense to pour the flaming brandy into the coffee from a height, risking accidents. Rather, choose the safer method and pour the coffee into the flaming brandy.

Pour the coffee into each of two decorative glasses. Finish with freshly whipped cream and insert a stirrer.

FLAMING COFFEE DRINKS IN THE GLASSES   Yield: 2 servings

Equipment

Réchaud or two-burner heat source
One chafing dish
One ladle
Dramatic tempered glassware for the finished coffee drinks
Gueridon

Ingredients

2 cups	Hot coffee
⅓ cup	Brown sugar
1 lump	White sugar
	Peel from half an orange, in one long strip
	Peel from half a lemon, in one long strip
2 or 3	Whole cloves
1 stick	Cinnamon
5 ounces	Cognac or other quality brandy

Procedure

Wheel the gueridon with its equipment and ingredients to the guests' table.

Keep the coffee warm in a thermal container on the side of the gueridon.

Light both burners and place one chafing dish atop a burner. Place 2 to 3 ounces of the cognac, brown sugar, lemon and orange peel, cloves, and cinnamon stick in the chafing dish and heat. Place one ounce of brandy in each of the tempered glasses.

Heat the glasses, with the brandy, by holding them by the stem and turning slowly above the heat. The rotation of the glass prevents the glass from overheating in any one spot (and breaking) and also helps to heat the cognac uniformly.

When the cognac in both goblets is sufficiently warm, tilt the cognac in one of the glasses just to the rim of the glass and allow the flame to ignite it.

Pour the cognac from one goblet to the other, holding one high and one low for effect. Transfer the flaming cognac back and forth, between goblets, a few times, then pour the contents of both into the chafing dish with the remainder of the warm cognac, igniting it.

Pour the coffee from the thermal coffeepot into the flaming cognac. Pour the coffee mixture into the two warmed glasses, top with freshly made whipped cream, and insert a stirrer.

PURCHASING AND CARING FOR AN ESPRESSO/CAPPUCCINO ❖ MACHINE

Most espresso/cappuccino machines are imported into the United States. In Europe, especially in Italy and France, espresso and cappuccino are very much in demand. The International Coffee Organization has estimated that approximately 1 percent of all coffee is served in the form of espresso, cappuccino, or one of their derivatives.

The first step in procuring an espresso/cappuccino machine is to determine which espresso/cappuccino beverages will be offered, how those drinks will be marketed, and what the estimated sales will be, especially during peak periods such as dinner and late evening.

Next, it is necessary to plan the layout of the facility and the placement of the machine(s). Is it better to have the machine(s) in the front of the house or the back of the house? Who will be making the espresso/cappuccino beverages—one person per shift or many bussers, servers, and/or captains?

What is the best way to convert the number of cups of espresso required into machine language? Remember that a "single group machine" fills two espresso (demitasse) cups simultaneously. Large "American" cups, however, are of such diameter that both streams of espresso will be directed into the same cup. If a guest has ordered caffè latte, which uses a large cup, a single group machine will produce one cup at a time. If two guests have ordered espresso, a single group machine will produce two cups of espresso simultaneously.

Consider the timing of making the cups of espresso and cappuccino. If a machine has an extraction time of 30 seconds (the time it takes the machine to push the steam through the grounds), one must figure making an espresso at a rate of one per minute. Assuming coffee beans are ground prior to service and are ready for the service staff, a staff member must locate a filter, fill it, pack it, smooth it off, then insert it into the holder, place and secure the holder in the machine, and turn the machine on or push the "go" button. The machine must begin to force the steam through the grounds, then allow the coffee to stream into the waiting cup(s). After making the espresso, the server should, ideally, remove the holder and dump the used coffee grounds and then rinse the filter, leaving it available for the next server.

Usually, however, the server approaching a traditional machine must remove the handle from the machine, remove the spent grounds, wash out the filter, then fill it again with fresh grounds, pack it, smooth it, and so on. Each step requires time. When figuring the rate at which a machine can produce cups of espresso and/ or cappuccino, include the server prep time before and after the production of each cup. If the servers will be steaming and frothing the milk for the cappuccino, that is an additional time-consuming step to be taken into consideration.

If two parties of four order espresso and cappuccino, an experienced server might take up to 8 minutes to prepare four cappuccinos using an extraction time of 30 seconds per espresso. Ten minutes from placing the order to receiving the cappuccino is an acceptable time frame for most guests. But what about the second party of four? That server must wait until the first party's beverages have been prepared, before beginning the espressos and cappuccinos for the second party. Now there is a minimum 20-minute wait until the cappuccino and espresso can be delivered to the guests. Is this exceptional service? Will this time frame be acceptable to either the guest or the establishment?

The preceding scenario assumed that the machine was connected to a water supply, a necessary arrangement for a commercial operation. Small machines must be manually filled and usually hold a gallon or two of water. A commercial establishment cannot take the risk of running out of hot water during service, having to refill the holding tank, wait for the water to heat, then continue with service. Not only is this an additional burden to the service staff (which will result in the staff not encouraging and even discouraging the purchase of espresso and cappuccino to guests), but the delay will inconvenience and irritate guests.

When deliberating about the machine to purchase (or lease), consider the basic types of machine. The traditional machines use either single- or double-service filters that must be hand packed with ground coffee, wiped off to produce a good seal, then fitted to the machine. To produce the espresso, the machine is manually turned on and off. All these motions require time and are subject to server interpretation and misuse of the machine.

More modern machines require the same manual preparation of the filter but the user merely presses a button to produce the espresso and does not need to turn off the machine once it is turned on.

Some espresso/cappuccino machines use filters that have been prepacked with the correct amount of ground coffee. These filters are set into the filter holder and espresso/cappuccino is made in the otherwise traditional method. This time-saving innovation also prevents the use of too much or too little coffee, thereby ensuring consistency in the final product.

Newer machines not only provide measured portions but are button operated. At the touch of a button, coffee can be ground, placed in a filter, packed, and steam forced through the grounds into the waiting cup.

The steaming and foaming of milk has always been a "difficult" area of traditional espresso/cappuccino service. If the milk is not sufficiently cold, it will not foam well. Once the milk is scalded, it will not foam well. With so many variables, it is difficult, especially for novice machine users, to consistently produce good foam to top cappuccino drinks. Imagine the frustration of being under pressure to produce wonderful, foam-topped cappuccinos for a table of eight!

Some of the newer models of espresso/cappuccino machines have a refrigeration unit that stores the milk for cappuccino drinks and, at the push of a button, portions out a serving of milk and automatically steams and foams it. These fully automated models are user-friendly to new service staff members as well as those who have been on staff, and make it possible to provide guests with what they've ordered consistently and rapidly, time and time again. These machines are faster and easier to use than the traditional models but are also more expensive initially. Owners and managers must weigh the cost against the time and expense of staff training on a more traditional model as well as their frustration when the finished product is less than perfect. Consider also that employees are unlikely to try to sell/upsell espresso and cappuccino if the drinks are difficult and time-consuming to make.

In the final analysis, it might be more cost-effective and profitable to purchase or lease the most up-to-date, self-sufficient, easiest-to-use machine the operation can afford. Through marketing and volume sales, the difference in base cost can be offset.

Whenever possible, purchase or lease a unit that has push-button service unless the restaurant's style and theme requires a "traditional" or "Old World" machine with a pull-handle to force the steam through the grounds. The modern machines, with the push of a button, allow the user to choose espresso, double espresso, cappuccino, or any of several other options. Choose an importer that will provide the level of service you require including staff training. Check the specifications: optimal pressure for the infusion of the steam through the grounds is 9 atmospheres. The optimal infusion temperature is 190°F. Both of these should remain constant throughout the production cycle. Consider the warranty period as well as service and repair options.

These are other questions that must be considered. Where a four-group machine will suffice for peak demand, would it be better to obtain two separate two-group machines, in different locations, to better service the guests and to ensure there will be a backup if a machine is out of order? How far away is the closest service dealer? Will the company leave replacement parts in-house for frequently replaced parts? Does it guarantee service within 24 hours? Remember, there are many parts on an espresso/cappuccino machine.

If the machine is to be in view of guests, the decor and design of the machine (e.g., the color of the side panels, whether it is brass- and copper-plated) become both aesthetic and budgetary concerns.

Finally, cups must be chosen and procured. The demitasse is the preferred cup for espresso, and a slightly larger cup is used for cappuccino. Regular American coffee cups are generally used for caffè latte or café au lait. Whatever cups are being used, they should be warmed prior to use. The espresso is infused at 190°F. To maintain its hot temperature, the cups ideally should be at 180°F to 185°F. Some manufacturers have a "cup warming" area atop the espresso machine. These areas generally can hold one to two dozen small cups. With any volume at all, however, a one- to two-dozen holding area will not be sufficient. In addition, if cups are brought from warewashing and stacked atop the cups already on the machine, the hot ones that should be used next will be on the bottom and the cooler ones, exposed to ambient air temperature, will be on top, the next ones to be used. Many of these machine-top cup warmers are only efficient and useful in low-volume operations or during nonpeak times.

A cup warmer should be installed near the espresso/cappuccino machine to hold the various sizes of cups required—typically demitasse, cappuccino, and American coffee cups. If the operation uses the American coffee cup for cappuccino, only two sizes of cups need be kept warm. The warmer should be set for 180 degrees Fahrenheit.

❖ MARKETING COFFEE DRINKS

Marketing of coffee drinks can be fun, interesting, and profitable. Some operations feature a late-afternoon "coffee/tea hour" and even invite local poets or authors for readings. Another idea is to provide entertainment for children (e.g., a puppet show) at one end of the dining room, while the parents and other adults, seated at tables in the back, sip the coffees and teas of the day.

During afternoons and late evenings, which are not peak service periods, coffee and tea service can utilize special decorative china—perhaps an oversized cup with the operation's logo. In addition, small edibles are often served, ranging from "tea sandwiches" to cookies. Generally, any cookie served during or with a traditional coffee and/or tea service is small and rather plain; it would not have chunks or bits of fruit, chocolate, or nuts. Favorites include lace cookies, madeleines, and flat butter, lemon, or ginger cookies. Today however, many establishments are creating new cookies, pastries, and other assorted goodies to serve with coffee.

Some of the more imaginative creations feature thin cookies with a filling such as a thin vanilla wafer spread with a hazelnut or chocolate filling, and then topped with another thin wafer. Or a layer of locally made jams may be spread between the layers.

Coffee cakes, from the traditional to the nouvelle, are often served with coffee and tea. The traditional coffee cakes range from rather plain cakes to those with crumb toppings, perhaps a prune filling and even topped with small, chopped nuts. Some establishments are now offering individual bundt cakes, spiraled with chocolate and nuts, and even containing a pudding center.

Freshly made *rugelach* is another European delicacy that can be exquisite with coffee and/or tea. They usually have various fillings of nuts, dried fruits, or bits of jam.

A novel idea would be to place a chocolate truffle on a separate plate, atop a doily, and serve it with the oversized cup of coffee or cup of tea. While not traditional, this treat is certain to appeal to those with a sweet tooth who will gladly devour the truffle then and there, or take it home for "later." In either case, it will be a difference maker that provides a reason for them to return to your operation as well as a positive experience to tell their friends about.

With breakfast or brunch service, individual 3-inch, heart-shaped waffles can be served. Later in the day, another batch of heart-shaped waffles can be made and dusted with powdered sugar and a slight dusting of cinnamon or chocolate shavings. They can be served with jams or fresh sliced fruits, such as bananas or peaches. If berries are in season, blueberries, raspberries, and the perennial favorite, strawberries, can be served with the waffles. Such a treat is imaginative and different, yet easy for the staff to prepare and serve.

Coffee-based pastries are often suggested to accompany coffee service, ranging from mocha truffles to Kona or Colombian coffee creams—chocolates with chocolate and coffee blended centers. Sometimes petits fours are created with coffee-based icings or fillings. Tiramisu, a traditional Italian dessert laced with espresso is another favorite of guests.

Chocolate and coffee are natural complements to each other. Chocolate cake is always popular; a chocolate torte with espresso frosting or fillings can add an elegant touch. Ice creams and low-fat and/or low-sugar yogurts and ice creams are also natural accompaniments to coffee service in flavors ranging from mocha to fruit (e.g., apple, cinnamon, chocolate).

If guests have time to linger and talk, perhaps a fondue would be appropriate. Try dried and fresh fruits (dried apricots, banana, apple, peach slices; and fresh strawberries and banana slices are favorites) with a hot chocolate or hot fudge sauce. For nonchocolate lovers, a crème anglais sauce or a caramel sauce would be wonderful.

Certain breads are also excellent with coffees. Usually, spicy breads are served such as apple bread, cinnamon bread, and the standby favorite, gingerbread. These can be either cubed or sliced or baked in cookie form. During holiday periods, certain traditional breads are favorites, including the 'Old World' stollen. These breads, which contain nuts and fruits, especially raisins and dried fruits, are dipped in rum and/or brandy while still warm, then brushed with melted butter and dusted with confectioner's sugar. Because these delicious treats are so rich, small portions are served.

And what better to serve with coffee than coffee beans—chocolate covered, that is? The enrobed coffee beans are available with dark chocolate, milk chocolate, and white chocolate; a mixture is especially attractive.

For those on low-fat diets, a piece of angel food cake or a small meringue, topped with fresh fruit, is perfect. Sometimes, meringues are shaped into tiny kisses or other decorative shapes, such as mushrooms.

❖   TEA SERVICE

Tea has been brewed for thousands of years. Various kinds of tea leaves create different types of tea. Manufacturers often add herbs, spices, and other flavorings to the dried tea leaves. Tea connoisseurs claim that the higher the elevation where the tea is grown (the higher the altitude), the better the tea.

Interestingly, the *Tea Importation Act* is the oldest existing statute that has been enacted to protect the purity of a food (tea sold in the United States). Tea leaves, being an organic product, can be easily contaminated. Microorganisms on the tea leaves could be harmful to humans and might not be killed during the brewing process. In addition, other substances such as dried herbs or grasses could be intermingled with the tea leaves without notice. Therefore, tea is inspected, regulated, and subject to controls and standards.

For centuries, tea has been subject to rituals. From the Japanese tea ceremony using a porcelain or ceramic pot of tea and handleless teacups to the formal English high afternoon tea service with silver teapot, creamer, and sugar bowl, and bone English china cups, rituals have developed in many civilizations for preparing, serving, and drinking tea.

The Chinese claim that Emperor Shen Nung accidentally discovered the making of tea approximately 4,000 years ago. According to the story, leaves fell into a pot of his drinking water, which was being boiled at the time. When the emperor tasted this flavored water, he found it to be refreshing, relaxing, and pleasant tasting. He wanted more. Thus tea was (supposedly) discovered.

The people of India, on the other hand, claim that a Buddhist priest, who needed to stay awake for an extended period, was searching for a stimulant and discovered tea. Supposedly he tried brewing various potions, foods, beverages, and finally tea. When he drank the tea, he realized that it gave him energy and helped him remain alert.

The *East India Trading Company* brought tea from China and Japan to Europe in the early 1600s. Queen Catherine, wife of King Charles II, was one of the first to receive a gift of tea. She and her husband tried it, liked it, and introduced it to their court and members of European society at afternoon tea parties.

Approximately 50 years later, tea was introduced to the American Colonies. It was very expensive but part of the English customs that were transported to the New World and made part of high society's life, at least until the Boston Tea Party.

The tea plant is an evergreen shrub, part of the *Camellia* or *Thea sinensis* family. It can reach heights of 15 to 30 feet but for commercial purposes is pruned and kept 3 to 5 feet tall so pickers can reach the leaves. Once dried, the tea loses three-fourths of its weight; it takes four pounds of tea leaves to make one pound of dried tea.

Varieties of Tea

There are over 3,000 varieties of tea. They are named for the regions where they are grown, where they are used, and who made them popular. Three of the most basic teas, *black, oolong,* and *green* teas, are made from leaves of the same bush. It is the processing, microclimate of the plants (soil, amount of sunshine, rain, etc.), and fertilization that create the differences. The main components of tea include:

♦ Tannin, which provides the body of the tea, some flavor, and some aroma.
♦ Oils, which provide aroma and flavor.
♦ Caffeine, which provides the stimulation.

There are decaffeinated teas on the market, made with herbs as well as true tea leaves. Caffeine-free herbal teas are made from such ingredients as chicory, rose hips, orange peel, mint, apples, cranberries, raspberries, and chamomile. Earl Grey is among the traditional teas available in decaffeinated form.

Of the traditional teas, *black tea* is the most familiar. It is made from partially dried leaves that are fully fermented before being toasted. *Oolong tea* is semifermented, and *green tea*, which is not very popular in the United States but which enjoys great popularity in the Far East, is an unfermented tea. The English and

---

POPULAR BLENDS OF TEA

Assam	Rich, smooth, full-bodied Indian tea.
Ceylon	Blend of black teas. Hearty, rich flavor.
Darjeeling	Blend of Indian high-altitude teas. Full-bodied taste. Choice of connoisseurs.
Earl Grey	Blend of Chinese and Indian teas. Delicate, unusual flavor.
English Breakfast	Blend of black teas. Full-bodied.
Jasmine	Oblong leaf tea from China. Scented, exotic, fragrant.
Orange Pekoe	Fragrant and smooth. Very popular blend of several Ceylon black teas. Standard tea served in tea bags in restaurants unless individual, foil-pack options are made available to guests.

Russians consume the most exported tea, followed by the Dutch and Americans. The characteristics of the most popular blends of tea are shown in the accompanying chart.

Brewing the Perfect Cup or Pot of Tea

When preparing tea, purists insist that a superior-tasting beverage results only if the water is freshly boiled for each guest starting with cold water. Accordingly, if the hot water is obtained from a coffee urn, the tea will not taste quite as wonderful as if prepared

---

PROCEDURE FOR PREPARING AND SERVING TEA

1. Pour hot water into both the individual teapots and the teacups, allowing them to heat.
2. Gather supplies: spoons, sugar, milk, lemon wedges.
3. Choose the loose tea or the tea bag(s) each guest has requested.

   Measure loose tea into a tea ball (holder) if one is used. Otherwise reserve the proper quantity in a small container until it is needed for brewing the tea.

   An individual tea bag can be placed on an underliner plate for each guest (e.g., a B&B plate). In a fast-food or family restaurant, where high turnover is a priority, the tea bag may be placed on the side of the saucer next to the teacup.

Lemon wedges, if requested, should be placed next to each tea bag, on the same B&B plate. In a fast-food or family restaurant, both the tea bag and lemon wedge can be placed on the side of the saucer.

4. Gather the number of saucers and teapot underliners needed.

5. Put the tea ball into the teapot after pouring out the hot water and add the boiling water. If using loose tea without a tea ball, place the tea in the bottom of the pot and pour boiling water over the leaves and allow the tea to steep. Maximum suggested steeping time is 7 minutes, which means the tea should be served to the guest within that time limit.

If using tea bags, fill each individual teapot with fresh boiling water.

6. Pour out the hot water from the cups and using a clean, dry napkin, quickly dry them. Stack the saucers, accompaniments, cups, and teapot(s) on a beverage tray and, holding the beverage tray in the left hand, approach the guest(s) at the table.

7. If there are women at the table who ordered tea, approach the eldest; otherwise approach the first man who ordered tea. Place a warmed cup on a saucer and place them, in one motion, at the guest's right side. If a spoon is needed, place it on the table linen, on the guest's right side, but to the inside of the cup and saucer. Place the other accompaniments (e.g., milk, sugar) on the table.

In a fast-food restaurant, the spoon may be placed on the side of the saucer with the tea bag and lemon wedge.

8. Place an individual teapot, filled with either plain hot water or hot steeping tea, on an underliner. In one motion, place the underliner and individual teapot to the right of the guest's teacup and saucer.

If a tea bag is being used, ask the guest, "Shall I pour?" This is the cue to the guest to place the tea bag in the cup so the server can pour hot water over the tea, beginning the steeping process.

The guest who desires to brew the tea him/herself will respond, "No, I'll take care of it," or something to that effect.

9. Move to the next person who requested tea. Serve women first, then the men at the table.

fresh for each guest, prepared as needed by heating a pot of freshly drawn cold water on a stovetop burner.

Special Requests from Tea Drinkers

Where loose tea is being used, a strainer is often served to each guest in addition to the items mentioned earlier. As the tea is poured from the teapot into the guest's teacup, most of the leaves will be at the bottom of the teapot. Some may float to the top, however, and to prevent those leaves from going into the guest's cup, the strainer is placed atop the cup and the tea is poured through the strainer. Naturally, a holder for the tea strainer must also be provided, or there will be no convenient place for the guest to put the strainer after pouring the tea.

Some guests may ask for an additional pot of hot water, especially those accustomed to making their tea in a European manner, that is, strongly brewed. It is a custom for some in Europe to brew extra–strong tea in the teapot and then dilute it to the desired drinking strength in the cup.

Tea drinkers who have ordered tea in an establishment that brews the tea in the teacup of each customer before bringing the cup to the guest might also ask for an additional pot of plain, hot water. Generally, these are fast-food locations where the servers place a tea bag in the guest's teacup, fill it with hot water in the kitchen or at the sidestand, then deliver the brewed tea to the guest. This tea might be too strong for some guests, who will then require additional plain, hot water and perhaps a clean (warmed) cup to dilute the beverage to their drinking preference.

Customs and tea service vary from country to country. In some Far Eastern countries, hot buttered tea is a favorite. Some tea drinkers in Greece or Turkey place a sugar cube in their mouth, then sip hot tea or Turkish coffee over the sugar cube, sweetening the brew as they drink it. Others may add a bit of vodka, rum, whiskey, or cognac to their tea.

Traditional Tea Service

Traditional tea service is very ceremonial and lends an air of sophistication to an afternoon refreshment. Many restaurants already have signature tea service, among them the *Plaza Hotel* and the *Helmsley Palace Hotel*, both in New York City.

At the Helmsley Palace, guests are ushered into a grand room decorated in gold overlays; a harpist plays softly in the background. There are a variety of loose teas to choose from, as well as an assortment of tea sandwiches. The service, the china, the ambience are all impeccable.

At the Plaza, the service is similar. The decor, while still glamorous, is a bit more festive and contains a bit less gold leaf. Both are "fun, happening, wonderful" places to enjoy an hour or so with a friend, enjoying high tea.

Alternative Tea Beverages

ICED ORANGE PEKOE TEA Yield: 3 quarts

8	Orange Pekoe tea bags or 6 Tbsp loose leaves
9	Oranges
2	Lemons
1/4 cup	Honey
1 quart	Ginger ale
2 quarts	Water

Slice one orange, reserving slices for garnish. Juice oranges and lemons. Cut rind of one orange into strips, removing the thick white pulp from the inside.

Prepare tea by pouring boiling water over tea bags or leaves, allowing it to steep for 10 minutes. Strain. Add honey and stir until it dissolves. Add the citrus juices. Cool.

When ready to serve, mix with chilled ginger ale and pour over ice into glasses. Garnish with orange slices.

Spiced Iced Tea: Stick whole cloves into the orange rind and add 2 cinnamon sticks and one whole allspice to tea while brewing.

Prepared Juice Method

8	Orange Pekoe tea bags or 6 Tbsp loose leaves
1	Orange
2 cups	Orange juice, fresh squeezed
1/2 cup	Lemon juice, fresh squeezed
1/4 cup	Honey
1 quart	Ginger ale

Prepare as previously described.

TEA FIZZ Yield: 8 servings

8	Tea bags or 6 Tbsps loose leaves
1 quart	Water
1 cup	Honey
11	Oranges
3	Lemons
1 quart	Ginger ale
1 quart	Soda water

Reserve one orange for garnish. Bring water to a boil and pour over tea bags or leaves. Steep for 10 minutes. Add honey and stir until dissolved. Juice oranges and lemons; add orange and lemon juice to tea.

To serve, pour tea mixture over ice cubes in glass to half full. Fill to three-fourths full with ginger ale; top with soda water. Garnish with orange slice.

FRUIT MEDLEY   Yield: 8 servings

8	Tea bags or 6 Tbsp loose leaves
1 quart	Water
1 cup	Honey
35	Oranges
8	Lemons
1	Pineapple
1 pint	Raspberries
1 quart	Soda water

Reserve one orange for garnish. Bring water to a boil and pour over tea bags or leaves. Add honey and stir until dissolved.

Steep for 10 minutes. Juice oranges and lemons. Peel, core, and juice pineapple. Puree raspberries and strain. Add fruit juices to tea.

To serve, pour tea mixture over ice cubes in a glass to three-fourths full; top with soda water. Garnish with pineapple slice or orange slice.

HOT SPICED TEA   Yield: 2 quarts

A favorite at ski resorts and ice skating arenas, or with anyone coming in out of the cold. Try it after shoveling snow, snowmobiling, or walking in the woods on a cold day.

1	Lemon
1	Orange
12	Cloves, whole
5	Allspice, whole
2	Cinnamon sticks
8	Orange Pekoe tea bags or 6 Tbsp loose leaves
2 quarts	Water
1/2 cup	Honey

Juice lemon and orange. Cut rind into sections, removing thick white pith from rind. In a saucepan, combine rinds and spices with one quart of water. Simmer for 15 minutes. Bring to a boil, pour over the tea bags or leaves, and allow to steep for 5 minutes. Boil remaining water. Add orange and lemon juices and honey to brewing tea; add additional quart of boiling water and steep for 5 additional minutes. Strain. Serve in warmed cups.

T e n

Extravagant Foods—
The Magic of Caviar

Extravagant foods can be a focal point, and a "fun" attraction for your establishment. Wherever an operation is located, there is usually some theme that can be capitalized on using an unusual or different food or food preparation—something that can become a signature dish, or a "fun, whimsical" giveaway or just a conversation piece, whether anyone orders it or not.

❖ MARKETING SUGGESTIONS FOR EXTRAVAGANT FOODS

The possibilities for extravagant foods are endless. In Louisiana or Florida, perhaps alligator tail meat can be featured. The lounge and restaurants and other food outlets might offer an appetizer of "'gator tail," cubed and either baked or breaded and deep fried and served with an assortment of hot sauces, steak sauces, and barbeque sauces. These could be offered as a happy hour giveaway or complimentary hors d'oeuvre once a month during the dinner meal period.

There should be a marketing trial for any such delicacies to decide whether they should go on the appetizer menu or late-night menu permanently. In the Southwest or other areas of the country where edible snakes are native, "snake bites," prepared in the same way as 'gator tails, might be a viable appetizer or late-night menu "attention getter."

Mr. Larry Leckart, owner and manager of Ronnie's Restaurant, once ordered an 8-ounce tin of beluga caviar for his restaurant. Understand that Ronnie's Restaurant serves quality food, in serious portions, at a reasonable price, but by no stretch of the imagination would it be regarded as an elegant, chic, or sophisticated restaurant.

Larry invited his regular customers to try the caviar—free of charge. It was Larry's way of thanking his regular customers for their patronage and providing something for them that, individually, very few, if any, of them would ever be able to afford to try. And they came back, again and again. Not necessarily for more caviar, but because they never knew what surprise Larry might have in store for them. And if there were no surprises, there was always great food, great big portions, and the standard Ronnie's service!

Upscale establishments might want to try caviar service as a regular, signature item; other restaurants might want to try it, as Larry did, for a limited time only. In either case, this chapter provides some tips about caviar.

❖ # THE MAKING OF CAVIAR

In processing caviar for the American and European market, the manufacturer carefully removes the roe, or fish eggs from the pregnant female fish and places the mass atop a screen. The size of the mesh of the screen will determine what size fish roe or berries (individual eggs are called berries) will be separated from the mass.

The berries are first processed through a rather large mesh to separate the biggest ones ("000" size, for large beluga caviar berries), then through a medium-size mesh to obtain the medium size ("00," for medium-size beluga berries), and finally through a rather tiny mesh for the smallest eggs ("0," for small beluga berries). While separating the berries by size, the mesh also separates out any membranes.

The berries that get damaged or crushed are not discarded; they are also valuable and marketable. The crushed berries are gathered (all types of sturgeon—beluga, osetra, and sevruga—may be combined) and prepared as "pressed caviar." Pressed caviar has all the flavor of whole berry caviar, but resembles more of a gooey jam than fine caviar. It can be spread on toast points or used in the same manner as regular, whole berry caviar.

The next step in the making of caviar is the salting. Salting refines and brings out the taste of the caviar. Most of us, if we ever had occasion to try plain, untreated caviar, would not like it. It needs to be enhanced, and salt is the ingredient to do just this. The addition of less than 5 percent salt results in malosol (or malossol) caviar. Thus "malosol" is not a brand or type of caviar, but designates certain criteria for the processing of the caviar, much as brut refers to a certain processing style, not a brand, of champagne.

Malosol (low-salt) caviar is the most highly regarded. Too little salt will not bring out the desired flavor; too much salt and it will taste too salty. Much as a winemaker must blend just the right varieties of grapes to produce a fine wine, and must allow it on its lees for just a certain amount of time, so must the caviar master determine, for each batch of roe, how much salt to add. The position of caviar master is sought after, and it takes years of experience and sensitive taste buds to become one.

Salt, besides enhancing flavor, can shrink the eggs, pulling the moisture out of them, and even alter the color. It is imperative that the caviar masters use only as much salt as necessary. One pound of salt treats approximately 36 pounds of caviar using the preferred malosol method.

Borax is often added to caviar for the Russian and European markets (it is illegal to add borax to caviar to be sold, imported, or otherwise marketed in the United States). Borax tightens and firms up the berries, and even sweetens them just a bit, which explains why borax is used in traditional Russian caviar processing and why Russians and many Europeans prefer caviar prepared that way. Borax however, can mask berries that are a bit soft or, in fact, on the verge of becoming unmarketable. Borax has been found to be harmful to humans and for these reasons was banned in caviar destined for the United States.

Additives, including salt and borax, are not indicated on tins of caviar. You must know the suppliers you are dealing with, their reputation, their integrity, and their past history in regard to claims they've made for their products. A person with a very good and experienced caviar palate can usually determine after purchasing and tasting the caviar whether it was treated with salt alone or salt and borax, but by then, it's too late—the container has been opened and it's being readied for service.

Packaging of caviar and temperature control are vital. The packaging should be as close to a vacuum as possible and the tins (the typical packaging) should be tightly sealed. Temperature must be carefully controlled; it is vital to maintain caviar at 26°F to 28°F. It won't freeze at these temperatures—the salt prevents freezing.

Caviar tins are often packed in ice both prior to and after opening. By surrounding the tins in ice, you can gauge the temperature quite accurately—if it gets too warm, the ice melts. This means the temperature is approaching 32°F and that more ice is needed to cool the caviar to the proper temperature of 26°F to 28°F.

Only the fish roe from beluga, osetra, and sevruga sturgeon is truly considered caviar. Due to the scarcity of sturgeon, however, many substitute fish roe have become popular. Generally, these are labeled with the type of fish as a modifier, (e.g., salmon caviar, paddlefish caviar, or lumpfish caviar). While not true caviar, these can be quite tasty and have their place among hors d'oeuvres, appetizers, and garnishes.

If someone fishing catches an alternative "caviar fish" (e.g., American sturgeon, salmon, or paddlefish), caviar can be made at home.

How to make your own "caviar"? The fish should first be dressed by cutting around the base of the tail and removing the cartilaginous skeleton with a twisting/pulling motion. Bleed the fish for about an hour, then carefully remove the roe. Meanwhile, procure a fine mesh or screen with holes slightly larger than the roe. Wash the roe atop this mesh or screen, gently stirring to separate the eggs from the membrane and allowing them to fall into a bowl below the mesh. Wash in cold water three or four times, draining the water quickly each time so the eggs are not saturated with water. Salt the roe, using 5 or 6 ounces of salt (fine grain, not kosher) to 10 pounds of eggs. After salting, enclose the roe in an airtight container. Fill the container to the top so there is minimal contact between the eggs and any air. Refrigerate at 26°F to 28°F. They will be ready to serve in one week. They will last 3 to 4 months if maintained at proper temperature with minimal air contact. When the eggs begin to get mushy and soft, it's time to discard them.

❖ TYPES OF CAVIAR

There are many types of fish roe, but very little caviar. As one caviar expert said, "Not liking lumpfish caviar does not mean that you don't like caviar." The most delicious and prized types of caviar are reviewed first, followed by more common "caviars."

Caspian Sea Sturgeon Caviar

The most famous, sought-after, and enjoyed genuine caviar is that of three Caspian Sea sturgeon. These three varieties provide over 80 percent of the world's true caviar. Sturgeon, a large fish that was in existence during prehistoric times, has survived through the ages, living primarily on plant life, with insect larvae, small crustaceans, mollusks, and an occasional fish supplementing its diet. Like the salmon, it lives in salt water but returns to fresh water to spawn. This pattern of returning to fresh water to spawn makes it an easy fish to catch in spite of a mature beluga sturgeon weighing over a ton. The sevruga and osetra sturgeon reach reproductive maturity in approximately 7 to 13 years, but the beluga needs 20 or so years. This makes the harvesting of the roe a critical undertaking so that the sturgeon does not reach the point of extinction.

Sturgeon roe vary in size and flavor depending on the species. The following sturgeon are prized for their roe.

Beluga

Beluga (buh-loo-gah) caviar, regarded as the finest quality caviar, comes from the *Huso huso* species of sturgeon found in the Caspian Sea. This variety is very rare. The berries should be large and firm, just slightly covered in their own oil, not soft nor mushy (which could indicate storage at improper temperatures, freezing and thawing, or other improper handling). The berries should be grey, varying in hue from light to very dark. There should be no hints of yellow, brown, or green—such colors indicate the eggs are from another species. This grey coloring of the berries—actually a lack of color—is a primary key to distinguishing beluga caviar from other caviars. The taste should be faintly salty (but not noticibly so) fresh-tasting, and not fishy—almost with a nutty flavor. The beluga eggs are graded from the largest (000) and lightest in grey coloring to smaller and medium grey in color (00), to the smallest and darkest (0). While there are differences in color and size, no one color is deemed "best," although the larger eggs are often prized due to their rarity.

The beluga is found primarily in the Caspian Sea, the largest inland body of water. The sturgeon roe is therefore primarily harvested and marketed by Iranians and those of Russian heritage. It is also found in the Sea of Azov and in the eastern Mediterranean.

The first or best quality of beluga caviar is processed as malosol, with little salt. The finest beluga berries are slightly softer than other sturgeon berries but are the largest and are still firm when they are fresh.

The traditional caviar tins from Russia have any of three colors on the cover. Beluga caviar has a blue circle bordering the outer portion of the top of the round 8-inch tin. The osetra has a yellow rim, and the sevruga a red rim. While the packaging may change, the traditional tins, if available, are decorative and "collectible."

Osetra

Osetra (aw-set-tra) caviar, from the *Acipenser sturio* sturgeon, is often regarded as the second most pricey caviar. Some connoisseurs regard it as the most tasty of the sturgeon caviars as, in addition to being rare, it has a uniquely zesty flavor, making it highly prized. This species is smaller than the beluga, growing to perhaps 600 or 700 pounds. The berry color varies from very dark greyish-brown-black to golden. The smoked fish meat (smoked sturgeon) is considered a delicacy. It is also spelled Osietr, Asetra, or Ossetra.

Sevruga

Sevruga (suh-vrou-gah) caviar is from the *Acipenser sevru* species, the smallest of the three Caspian Sea sturgeons. When mature, it weighs approximately 40 to 50 pounds. The sevruga has, among the beluga, osetra, and itself, the smallest eggs, ranging in color from light to dark grey, almost black. Some connoisseurs consider the sevruga the best price performer of the three Caspian Sea caviars, providing nearly as good a taste as the beluga for much less cost.

Pressed Caviar

Pressed caviar consists of processed mature sturgeon berries that are slightly damaged and/or crushed. The resulting "pressed caviar" resembles a gooey spread but can be quite delicious. It is often served atop blintzes with sour cream or used for other decorative, garnishing purposes.

Chinese Caviar

A relative newcomer on the caviar market, the "Chinese beluga" comes from the *Huso dauricus* species of sturgeon. It is often as pricey as sevruga and sometimes as much as osetra with berries varying from very dark grey-green to a brownish-yellow.

Chinese caviar is also called "Kaluga caviar," "Manchurian caviar," or "Mandarin beluga caviar." It is generally from the Amur River and is a good price performer, although the key is to know the supplier as the consistency must be watched from shipment to shipment. The best of the Chinese caviar is processed as malosol caviar. The eggs vary in color from a dark grey-greenish-brown to a yellow-green-brown.

American Caviar

These caviars are very close in price and taste to sevruga caviar. Several species of sturgeon and other fish produce "American caviar," including the paddlefish or spoonbill (Mississippi River area),

Atlantic sea sturgeon, white sturgeon (Pacific northwest), hackleback or shovelnose (Mississippi River area), sturgeon, and the bowfin. Most of the varieties of the American sturgeon take approximately 7 years to reach sexual maturity, similar to the sevruga.

The *hackleback* or *shovelnose sturgeon*, generally found in the Mississippi River area, produces some of the best American sturgeon caviar. It looks like black lumpfish caviar, with small jet black berries, but has firm berry texture and real caviar flavor, almost a sweet nuttiness.

The *paddlefish* (so named for its paddlelike snout) grows to about 200 pounds, producing both roe and smoked fish that are highly marketable. While not a true sturgeon, the roe is valued for its excellent quality. Paddlefish caviar resembles the Russian sevruga. The berries vary from a dark grey/black to golden in color. Paddlefish caviar is considered to be an excellent American caviar.

The *American white sturgeon* is found in waters of the Northwest. It produces excellent "white sturgeon caviar," but it is limited in quantity.

The *bowfin*, or *choupique*, is not a true sturgeon, but is a good lesser quality yet definitely affordable American caviar. It is viewed as better than lumpfish caviar although it can, at times, taste "fishy." The berries are dark brown to black in color. Bowfin caviar is distinctive because on heating, the berries turn bright red, looking like a smaller version of the salmon caviar.

Salmon Caviar is obtained from the Pacific salmon, mostly from Alaska. This caviar can be close to sturgeon caviar in quality. Due to the berries' size and color, there is quite a high level of demand for salmon caviar—the berries are large, firm, and coral-red to orange-yellow in color, making it truly unique and a lot of fun to use as a garnish.

The berries are prepared for the American and European markets in a manner similar to that of sturgeon roe; there is a different preparation for the Japanese market. Generally the pink, chum, and coho salmon roe are used for salmon caviar (chinook salmon eggs are generally too large to make good salmon caviar). The berries are separated from connective tissue using a screen mesh, then immersed in a brine solution for 15 to 30 minutes until the berry coagulates (firms up) but does not begin shrinking. The brine solution will draw the moisture from the berries if they remain in it too long. The berries are then drained, packed, and refrigerated.

Whitefish Caviar

The *Coregonus albula* species of whitefish is known chiefly for its roe, rather than for the flesh of the fish. This caviar is also known as *Vendance* caviar. Whitefish caviar is from European freshwaters, Britain to Scandanavia to Siberia. It is pink in color and each berry is so tiny that the caviar has a pastelike consistency. It is considered excellent in taste and texture.

There is another whitefish caviar from fish from the Great Lakes. This whitefish caviar is yellow in color and small in size, similar to the sevruga. It is more readily available to the general public and is a wonderful addition for hors d'oeuvres, to use as a garnish or with appetizers. In general, it is great for use with food but not really a good choice to serve alone, as with beluga or osetra, or sevruga caviar. It is usually labeled "golden whitefish caviar." It is crunchy in texture, rather than firm and sometimes is dyed red or black.

Lumpfish Caviar

The lumpfish is found on both sides of the Atlantic. Each reproductive female produces a vast number of very small (^{1}/$_{10}$ of an inch in diameter—the size of a pinhead) dark green to yellow eggs. Most commercial lumpfish caviar is produced in Iceland. It is usually marketed as a red or black "lumpfish caviar," having been dyed. The quality of lumpfish caviar is inversely proportional to the amount of salt used. Scandanavian mildly salted lumpfish roe is much more valued than the commercial lumpfish caviar, which is, generally speaking, overpreserved (oversalted). Commercial lumpfish caviar is generally used as a garnish where just a few small berries are used on each appetizer or hors d'oeuvre. It is used more for the illusion of caviar than the taste of caviar, having more color and texture than any (other than salty) taste. The least expensive of the substitute caviars, it is often pasteurized and marketed in small glass jars that can remain on store shelves indefinitely. Definitely not a true caviar, but it has its place as a garnish.

Other "Caviar"

With the decline in the sturgeon population—especially the beluga, osetra, and sevruga sturgeons of the Caspian Sea, and the American sea sturgeons—caviar lovers have had to turn to alternatives.

In addition to those previously mentioned, the following are possible substitutes:

- *Alewife*, or *Herring Caviar.*
- *Carp Caviar.*
- *Cod Caviar*, or *Hon-tarako* in Japan. Usually dyed red.

♦ *Oeufs d'Escargot,* or *Escargot Caviar.* From the Petit Gris Helix Escargot. The eggs are the size of beluga or sevruga and have an "earthy" flavor.

♦ *Lobster Roe,* or *"Coral."* Possessing a coral color, this is often used in sauces to accompany lobster dishes, but can be used alone or as a garnish.

♦ *Walleye Pollock Caviar,* or *Alaska Pollock Caviar.* A member of the cod family. Roe are salted and dyed red, sold as *Momijiko* caviar.

♦ *Mullet Caviar.* From the Florida mullet, the roe is a delicacy in Florida. Often served fried with scrambled eggs or as a garnish on heart-of-palm salad. When salted and dried, it is known as *karasumi* in Japan; *bottarga* is a similar mullet caviar found in Italy; *batrakh* is a similar mullet caviar served in the Middle East.

❖ STORAGE OF CAVIAR

Caviar should be kept well chilled at all times. Storage temperature should range from 26°F to 28°F. Temperatures higher than 30°F will cause the quality to deteriorate rapidly.

Freshness is the number one criterion for purchasing caviar. Fresh malosol sturgeon caviar will be superior in taste, texture, and color. If the caviar is purchased in a vacuum-sealed container and is kept adequately chilled, it may be kept for up to 6 months. Once opened, however, it should be used quickly (within five days) for maximum flavor and freshness.

Sturgeon caviar should not be frozen as the berries will burst from the expansion of freezing and become soggy upon thawing. The texture of the firm berries will have been ruined and the taste will deteriorate. Malosol sturgeon caviar will freeze below 26°F, so be sure not to store it below that temperature. However, salmon caviar and whitefish caviar can be frozen without harming the berries.

Caviar should be kept in the tin or jar it was shipped in with minimal contact with air and water minimized. If a portion of a tin is used, repack the remaining caviar in a smaller glass jar with a tight-fitting lid. Try to minimize the air pocket at the top of the jar. If the caviar has not been opened, rotate it at least twice per week to distribute the caviar oils evenly over the berries.

When purchasing caviar, try caviar samples from the supplier before purchasing an 8-ounce or 10-ounce tin. Look for whole, uncrushed berries with a shimmer of oil on them. Some might have been crushed in the processing, but at least 95 percent of the berries should be individual and whole. The caviar should smell fresh, not fishy or salty. If there is any aroma, it should be almost a "seabreeze" scent. Look for the characteristic color and size of berry for the type of caviar being purchased.

❖ # SERVICE OF CAVIAR

Caviar service could be an adjunct to high tea service, or a specialty in its own right. Traditionally, during caviar tastings, hot unsweetened tea is used to cleanse the palate. Caviar, accompanied by such tea, or by the very popular brut Champagne or shots of ice-cold vodka might become a feature specialty of your establishment.

When using caviar tasting as a focal point, or point of difference, between your establishment and others, you could start off with a Champagne "chaser," preferably a brut so the bare minimum of sweetness will complement the taste of the caviar, neither losing nor overpowering its flavor. The wine should be brought to the tasting tables in ice buckets, to ensure it remains very cold, and should be served in champagne flutes. Vodka, another traditional accompaniment to caviar tastings, could be served in small, individual-size carafes that are immersed in ice shavings and so kept chilled. If there are several people at a table, several individual-size carafes can be inserted into a decorative dish of shaved ice and placed at each table. If hot unsweetened tea is the liquid accompaniment to the caviar tasting, the basic tea service outlined in the previous chapter would be followed. For a caviar tasting, however, often pots of hot tea will be brewed and poured directly into individual teacups so the caviar will remain the focal point of the gathering.

Present caviar regally. Silver and silverplate should never touch caviar as it will have two distinctly bad effects. First, the caviar will have its flavor significantly altered for the worse, and second, the caviar will stain the silver or silverplate. Stainless steel is not a good alternative, either.

Caviar should be presented in a glass jar or bowl or the original tin or glass jar, and should be embedded in ice. The glass or tin might be placed in a silverplated or decorative shell-shaped bowl to create the illusion of elegance so befitting caviar. The serving piece would then be heaped with ice, around the caviar container to keep it well chilled.

In other words, the caviar presentation should be "layered," with the silver tray or shallow bowl on the bottom, heaped with chipped or shaved ice. The ice supports and chills the caviar which is contained in the glass jar, bowl, or tin.

Utensils to serve caviar should likewise not be silver, silverplate, or stainless steel. Serving pieces should be carved mother-of-pearl spoons, or as a second best, plastic.

Sturgeon caviar usually should be served with only toast points and either brut Champagne or ice-cold vodka. For a tasting or during high tea, the caviar may be served with unsweetened hot tea as well as Champagne or vodka. Some people spread a bit of unsalted butter on crustless bread before toasting it. Purists, however, demand the plain toasted triangles.

American and Chinese caviars are sufficiently high in quality and resemble true Caspian Sea sturgeon caviar closely enough that they might also be presented in the preceding manner.

Other caviars, however, are best suited to garnishes, hors d'oeuvres, and "fun" preparations. Typically, the "caviar" is placed in the middle of the table surrounded by a variety of accompaniments:

◆ Finely chopped hard-boiled egg whites.
◆ Hard-boiled egg yolks pushed through a sieve.
◆ Finely chopped onion.
◆ Sour cream.
◆ Lemon juice (if very heavily salted).
◆ Chopped fresh chives.

If serving pressed caviar, try one of these fun presentations. Heap it on warm blintzes, cover with chopped onions, chopped hard-boiled egg whites, and sieved yolks, and finally top each with sour cream. Or hollow out cooked red potatoes and fill them with pressed caviar, chopped onions, chopped fresh chives, and finely chopped hard-boiled eggs; then top the potatoes with sour cream. Both variations are wonderful!

Another traditional gourmet presentation is to top sturgeon caviar with minced, cooked pheasant breast that has been chilled. The caviar is placed atop the traditional toast point, then a bit of minced (by hand with a chef knife) cold pheasant breast is drizzled over the caviar.

French chefs have long taken squares of thin smoked Norweigan salmon, topped them with beluga, osetra, or sevruga caviar, then rolled the salmon into cone shapes or cornets.

Czar Nicolas, it seems, liked his oysters sprinkled with chopped onion and beluga caviar. Definitely worth a try!

The key to serving caviar is to highlight it if it's sturgeon caviar or other good quality caviar, and to have fun with it and just enjoy it if it's a lesser quality caviar.

Because of caviar's high cost and perishability, it commands a top price and top billing on the menu. It is viewed as the perfect "celebration" food, especially since it is often paired with that popular celebration beverage, Champagne, for those special occasions guests want to remember.

❖ DETECTING LESSER GRADES OF CAVIAR

The most-often perpetuated fraud in the sale of caviar, both to the restaurateur and to the public is substitution of species. How can you determine whether what you think you have purchased, is what you really have?

The best way is to know your supplier and to know the restaurant where you are having the caviar, but even that is not foolproof.

A true caviar connoisseur will be able to identify the type of caviar from the berry oil. While very few of us have the caviar-tasting experience and palate to be adept at identifying caviar oils, the following indicators are reliable:

◆ If it is beluga caviar, there will be no trace of color, only shades of black and/or grey. If there are traces of yellow or brownish-green, the berries are not beluga, but possibly berries of another species, often osetra.

◆ Apply heat to a few berries—put them under the broiler, in the microwave, or in the oven for a few minutes. If the berries turn red, they are bowfin, not sturgeon.

Rare Wines

Many restaurants pride themselves on maintaining a selection of vary rare wines for sale to their guests, dating from perhaps an 1865 Lafite Rothschild to an 1865 Romanee-St. Vivant (market/auction prices averaging $550) to a 1928 Latour (about $1,823 at auction) to a 1947 Petrus ($4,458 at auction) (prices approximate). Why? Among many reasons, the following are the most likely:

♦ Rare wines can be a conversation piece. Guests take one look at the wine list and exclaim something on the order of, "Dear, look at this . . . who in the world would spend $[XXXX] on a bottle of wine?" And on and on, they go. And guess what happens when they get home? They begin talking with their neighbors, exclaiming something to the effect of, "Why, did you know that at Z Restaurant they have bottles of wine that sell for $[XXXX]. Can you just imagine! Who would ever buy such a bottle?" And you've just received wonderful word-of-mouth advertising. Not only will the people who dined there remember your restaurant, but the neighbors will, too. And, as long as the dinner was tasty and the service excellent, they now have something fascinating and different to come back for—another look at your wonderful wine list, whether or not they ever purchase a bottle of the rare wines.

♦ There are those who really do purchase and drink rare wines. Some are connoisseurs who might have their own wine collection in their home wine cellar but who are in your city on business or vacation and happen to be visiting your restaurant. If you have a wine that they'd enjoy, they'll order it and not mind paying $300 or $400 or even $1,000 for a great Burgundy or Bordeaux. Or perhaps the guests live in your geographic area but enjoy trying different rare wines. Knowing of your selection of rare wines and reputation for maintaining and storing them properly, they may choose to dine at your restaurant just to try some of the wine selections that they don't have in their personal wine cellar.

♦ There are the special occasions. Someone is getting engaged or married, and they want only "the best." Someone just closed a lucrative real estate deal; the commission on the deal is in the six figures, and they want to mark the occasion. A stockbroker just sold a couple million shares of stock and the commission is, again, in the six figures—they want to celebrate. A sports figure just signed a megamillion-dollar contract. An entertainer is in town performing and stops by before or after the show. A classic way to celebrate is to order a wine they've always wanted to try but felt would be too great an extravagance except for that "special occasion." If it's not on your wine list, they can't order it.

♦ There are foreign guests. Exchange rates for various currencies are always in a state of flux, but throughout the past decade, visitors from many foreign countries have had the benefit of the exchange ratios. A wine for $400 may seem like a bargain to them, or at least, a "normal" price for a bottle of good wine. If such wines are on your wine list, they can be sold.

♦ And there is always the lottery winner. The lucky winner from the casinos in Las Vegas or Atlantic City, or the state lottery winner. Improbable, you say? Yes. But it could happen, and if you don't have the wines on your wine list, you can't make the sale.

❖ HANDLING RARE WINES

Rare wines are delicate and should be preserved in a temperature-controlled and humidity-controlled storage cabinet or area. Many commercially available "closets" offer temperature and humidity control; if the building has a cellar, it can often be converted quite easily to accommodate rare wines. Fine wines should be maintained at a constant temperature of 50°F to 60°F for aging (for serving, white wines should be cooled to 40°F to 47°F, and red wines to 60°F to 68°F). Humidity should range between 55 and 75 percent. The wine cellar should be free from vibration—a refrigerator will not suffice, even for white wines. The motor will cause the wines to vibrate so they will be in a constant state of being "shook up"; they will mature faster than if kept undisturbed in a cool, humidity-controlled, still environment.

If the wines are permitted to be in a warmer-than-desirable environment, they will expand, possibly forcing wine out around the cork if there is room for leakage. When the wine later cools, the cork and possibly some air may be pulled back into the bottle. Sometimes when this happens, only air is pulled back into the bottle and the cork remains pushed out a bit. To the trained eye, this is a sure sign of improper storage.

Rare wines need to be handled very gingerly. Sediment accumulates within the bottle as the wine ages and slowly drifts to the bottom of the bottle. If the wine is stored on its side, as it should be to keep the cork moist, the "bottom" actually will be the side it is resting on. The cork needs to be kept moist to prevent shrinkage. If the cork dries out and begins to shrink, it will pull away from the sides of the bottle and allow air to enter. Air is an enemy of wine; it will cause the wine to mature at a much faster rate than if properly stored. The wine after much exposure to air will become oxidized or spoiled and may taste vinegary.

Cork is a natural product, made from the bark of the cork tree, and is used to close most if not all fine bottles of wine. Cork is the traditional closure dating from the days of Dom Perignon and early champagne makers. Some manufacturers are using artificial corks made of plastic, but these are generally used with "drinkable-when-made" wines, not wines to be stored for years, although this may change in the future. Due to constant expansion and contraction caused by changes in temperature and humidity, a cork can dry out, lose its elasticity, and pull away from the sides of the bottle. Over time, corks will dry out even if the bottles are stored properly.

Because of this tendency of corks to dry out and allow air to enter the bottle, rare and fine cellared wines should be recorked every 20 to 25 years. To properly recork a bottle, it should be carefully packaged and insulated as much as possible from jostling and shock, then sent to the original château or wine cellar where it was bottled. During the recorking process, the château tops off the wine by adding a bit of wine from the same year as the bottle. The "better" châteaus reserve a bit of each year's wine for "recorking and topping" purposes. Topping off the bottle brings the liquid level back either to a high fill (just above the shoulder) or to a top-shoulder fill. The bottle is then recorked, sealed, and returned to the sender. The "high fill" (the wine reaches into the neck area of the bottle) is the normal level for a new bottle of Bordeaux.

❖ ULLAGE OR WINE LEVELS FOR BOTTLES OF BORDEAUX

A way to gauge whether wine is still good or has spoiled is to check the *ullage,* or air gap, between the cork and the wine in a bottle.

The shoulder, or sloping portion of a bottle of wine, is generally used as the indicator of the "fill" level, especially in Bordeaux wines and wines bottled in Bordeaux-style bottles. These bottles are characterized by straight sides with a relatively short shoulder and a straight neck (See Figure 11.1 for an illustration of a Bordeaux bottle); burgundy-style bottles have a longer, sloping shoulder, or a longer portion of the bottle with sloping sides.

A Bordeaux bottle has various levels of fill. Both *Christie's* and *Sotheby's*, which are internationally renowned auction houses noted for their auctions of rare wines and their expertise regarding rare wines, define eight levels of ullage on a Bordeaux bottle. Figure 11.1 shows a sketch of the necks of Bordeaux bottles indicating various ullage levels. See Appendix B for drawings of Sotheby's and Christie's ullage levels and descriptions of each level.

Various ullage levels, or the level of wine in a (Bordeaux) bottle, are described as in Sotheby's catalog (S) and Christie's catalog (C). Regarding wines and their condition, due to corks or otherwise, *Christie's* catalog of April 9, 1992, states:

> Though every effort is made to describe or measure the levels of older vintages, corks over 20 years begin to lose their elasticity and levels *can* change between cataloguing and sale. Old corks have also been known to fail during or after shipment. . . . We therefore repeat that there is always a risk of cork failure with old wines and due allowance must be made for this. (p. 2)

The following are the ullage levels used by these auction houses: Please refer to Appendix B for illustrations of each level.

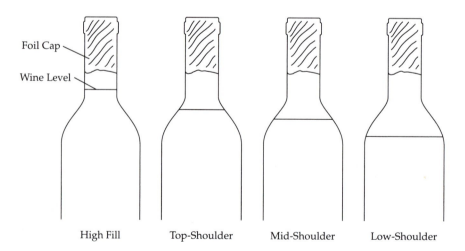

Foil Cap
Wine Level

| High Fill | Top-Shoulder | Mid-Shoulder | Low-Shoulder |

FIGURE 11.1   Sample ullage levels, Bordeaux bottles of wine.

1. A *high fill* (C) or *within the neck* (S) bottle of Bordeaux is one where the wine extends into the neck of the bottle. This is typical of young wines.

2. *Bottom neck* (S) or *into neck* (C) is acceptable for a bottle of any age and particularly good for a bottle that has been in the bottle for 10 years or more. The easiest way to describe this level is that it is the point at the center of the curve where the neck meets the sloping shoulder; it is neither neck nor shoulder, but the point where they meet.

3. *Very top shoulder* (S) is a level whereby the wine is just *barely below the midpoint of the curve* where the neck and shoulder meet; it is the highest point of the shoulder. This level is also acceptable for any bottle of wine.

4. *Top shoulder* (S, C) represents the point in a bottle where the wine reaches the top of the shoulder, but does not extend into the neck. It is the point at the *end of the straight segment of the shoulder*, not extending into the curved portion of the bottle where the shoulder meets the neck of the bottle. This is the usual level for (claret) wines 15 years or older.

5. *High shoulder* (S) or *upper shoulder* (C) is the level where the wine is above the mid-point of the shoulder (half-way along the straight segment of the shoulder) but below top-shoulder. This level may represent typical and natural reduction of the wine level through the cork and evaporation through the cork and capsule. Usually it represents no problem. This level is acceptable for wines over 20 years old and exceptional for pre-1940 wines.

6. *Mid shoulder* (S, C) usually suggests some deterioration and/or weakening of the cork and possibly some variation in taste, aroma, and bouquet from what would normally be expected for that wine. It is not abnormal for wines 30 to 40 years old. This is the point halfway between the uppermost point of the straight side of the bottle and the lowest point of the straight side of the neck.

7. *Low shoulder* (S) or *Lower mid-shoulder* (C) level is the level above the point where the lower bottle begins to curve into the shoulder section but below the middle of the straight segment of the shoulder. This level represents some risk in the quality of the wine in regard to its flavor, bouquet, aroma, and usual characteristics.

8. *Bottom shoulder* (S) or *Low shoulder* (C) is the point at the bottom of the shoulder where the shoulder meets the straight side of the wine bottle—the uppermost point of the straight side of the Bordeaux bottle. It is not usually sold for consumption but rather because it is an exceptionally rare or interesting wine.

For Burgundy and Hock (German) wines, which are packaged in bottles with long sloping sides, another measure is required. Distance, as measured in centimeters, is the standard measure. Christie's, in its catalog of April 9, 1992, states:

> Because of the slope of shoulder it is impractical to describe levels as mid-shoulder, etc. Wherever appropriate the level between cork and wine will be measured and catalogued in centimeters. (p. 2)

Sotheby's uses the centimeter measure for Burgundy and German wines as well.

Wines with a fill level at mid-shoulder or below are generally considered risky for consumption. Roughly two-thirds of the bottles with this level of fill are spoiled, or oxidized, and not suggested for consumption. If the wine is a 1960 vintage or more recent, it should definitely have a high or top-shoulder fill. If not, "buyer beware," as the saying goes.

Burgundies and German wines, with the longer sloping sides of their bottles, are more difficult to judge. Generally speaking, if the bottle is over 50 years old and has ullage of 1½ to 2 inches, it should be excellent; if the ullage is 2 to 3 inches, that distance is normal for a bottle of that age. Over 3 inches of ullage begins to get risky for the quality of wine in a Burgundy-shaped bottle.

Today, some winemakers are beginning to wonder whether corks are the optimal closure for a fine wine. In addition to drying out and crumbling with time, cork can deteriorate in other ways. It is estimated that 4 to 8 percent of wines bottled with corks have a "bad," "moldy," or "corky" smell when opened and cannot be consumed.

Since all corks dry out over time, but not at the same rate, there are necessarily variations in individual bottles of wine even when they are from the same vineyard, the same vintage, the same bottling. The corks become a variable to consider, just as storage conditions are a consideration. Two bottles of identical wine may taste very different simply due to the difference in the corks. Winemakers take great pains to ensure their wine is made properly; then they cork it and age it. Years later, when the wine should be in its prime, many bottles are flawed by dried-out, crumbly, leaky corks. Should this be the case? Is it necessarily the case? Is it time to consider change? In the years to come, purists and traditionalists will battle this question against those advocating change. Over the next decade or two, it will be interesting to see whether the wine industry changes its custom of utilizing cork closures on bottles of fine wines meant to age for 20, 30, or 40 or more years.

What should be done if a bottle, just removed and opened from a sealed (unopened) case, is discovered to be "corky?" Most reputable vintners, suppliers, and distributors suggest that they be contacted. The remainder of that case of wine will be checked, and, as long as the ullage or corkiness was not due to the collector's (or restaurateur's) storage, most vintners or suppliers will rectify the situation. Rectifying the situation usually consists of replacing the wine with new cases of the same wine, or the same wine but a different, comparable vintage, or a comparable wine. A wine deemed to be comparable, in this instance, is one that meets both the criteria of the organization doing the replacing and the organization asking for the replacement. This policy, according to most suppliers, applies only to purchases of closed (sealed) cases of wine. Single bottles or open cases of wine should be checked prior to purchase. If the ullage seems to be a problem, don't purchase the bottle.

SELLING AND SERVING RARE WINES

When selling rare wines to guests in a restaurant, it is important to stress this cardinal rule: *Rare wines are sold as is.* Often there is a special "Rare Wine List" or a section on the regular wine list where the rare wines are listed. Where the rare wines are listed, a disclaimer should appear: *"All rare wines are sold as is."*

After taking the order for a rare wine, the server should advise the guest, "Rare wine bottles of wine are *sold, once opened, as is."* Whatever condition the wine is, once opened, it is theirs . . . good or bad, drinkable or not. If a guest orders a bottle of rare wine, and, on presentation, notices that the ullage is below what he or she believes to be drinkable and does not accept the bottle, that is fine. The house accepts the bottle back and there is no charge. Once the bottle has been opened, however, regardless of whether the wine is drinkable or not, it is the guest's property.

Sediment in Rare Wines

It is equally important to stress to those who will be handling rare wines that it is necessary to keep the bottles as still as possible whenever moving them to avoid disturbing the sediment on the bottom. Moving the bottle too quickly or changing the direction of the bottle or moving it too vigorously will redistribute the sediment throughout the whole bottle, spoiling it for the moment. Before that bottle can be opened for consumption, it may take up to 3 days for the shaken sediment to fall to the bottom again. This is why it is important not to shake, move, rock, or otherwise disturb the contents of a bottle of rare wine while carrying it.

An analogy is to think of the sediment as a sort of "dust," very light and airy. Once it is distributed throughout the wine, it takes days to resettle. It is essential to avoid moving a bottle with sediment more than absolutely necessary, especially to change directions (horizontal to vertical).

Red wines, due to their age and the organic matter in the wine (yeast) that has been maturing, will probably have sediment. Never shake or move the bottle more than necessary. Except for the need for gentle handling and extra care in removing the cork, white wines and champagnes are treated in the same way as white wines and champagnes on the regular wine list.

If the house has "repeat" patrons who regularly drink rare wines, ask them to request the rare wines they anticipate ordering (and drinking) 1 to 3 days in advance so the selected bottles can be gently stood upright and the sediment allowed to settle. The sediment in a bottle lying on its side in rare wine storage will be deposited in a line running along the cylinder portion of the bottle. When sediment is deposited along the side in this way, it makes it much more difficult to open the bottle without disturbing the sediment.

The Choice of Corkscrew

Because old corks are brittle and crumbly, the choice of opener is an important decision. A "waiter's corkscrew," or an "Ah-so" cork puller, can be used for rare wines.

Collectors might acquire several corkscrews, each for a particular kind of cork. If it is well designed, the traditional corkscrew, a wormlike screw fitted into a handle of wood or metal, or a professional server's combination penknife, worm, and lever works well for most corks. The worm should be long enough to extend through the cork. Since some wines have long corks, a worm length of at least $2^1/2$ inches is recommended. The open, perfectly round wire helix, with an off-center point, a hollow core of about $1/8$ inch, an outside diameter of $3/8$ inch, and six complete spirals works best.

Old corks are easy to remove using a tweezerlike cork extractor, commonly called an "Ah-so," that does not penetrate the cork but breaks its seal on the side of the bottle. The needle extractor, which injects gas or compressed air under the cork also works well for old corks.

Very young or unseasoned corks require some strength to remove. Double-helix corkscrews, which facilitate a cork-wrenching twist, are useful. A good worm screw in a double-handled lever device, a professional wine waiter's lever penknife, and a screw-action or a screw-and-bell device also facilitate removal.

Glassware for Rare Wines

Rare wines may be served in special stemware or, on request, tasting glasses. If the establishment has special "rare wine stemware" it should be brought to the table as soon as the wine is ordered.

Tasting glasses are normally brought only on request as they are unique in shape, without a stem and of an odd appearance, having a flat bottom with a "bubble" rising in the middle of it and with sloping sides closing in at the top (see Figure 11.2). They are somewhat difficult to drink from. This shape is supposed to be the best for gathering the aromas and bouquet of a fine wine and will show off a wine's faults equally as well as its strengths. It is understandable why an establishment would not want to offer these glasses to everyone, and equally understandable why only connoisseurs of wine prefer to taste wine from them and request them.

The server sets the appropriate special stemware, if the establishment has such, on the table. Some establishments will have separate glasses for:

◆ Young red wines (less than 5 years old) and ports.
◆ White wines.
◆ Older red wines (older than 5 years) and older ports.
◆ Tastings, by connoisseurs.

Traditional shapes of wineglasses are shown in Figure 11.3.

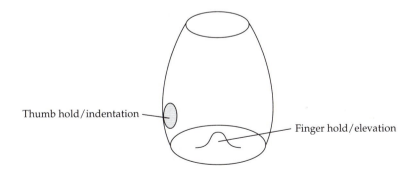

Thumb hold/indentation

Finger hold/elevation

FIGURE 11.2   Tasting glass.

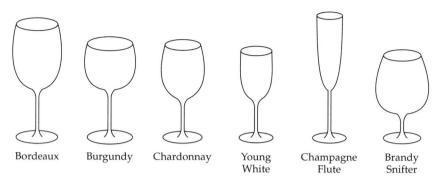

Bordeaux   Burgundy   Chardonnay   Young White   Champagne Flute   Brandy Snifter

FIGURE 11.3   Traditional shapes of wineglasses.

The Decanting Question

Should or should not a wine be decanted? This is an age-old question with various connoisseurs declaring that such-and-such a wine requires decanting while others don't.

Decanting can be a source of great ceremony at the table or in the presence of guests, or can be quite efficient and forthright if performed back of the house. Its purpose is to (1) separate the clear wine from the sediment in the bottle, and (2) aerate the wine, which releases the bouquet.

Whether the wine needs to be decanted to separate the wine from the sediment is a tricky enough question; whether a given wine requires aeration, and if so, for how long, is an even more nebulous consideration. Aeration helps to soften many wines, especially young wines that are harsh (lots of tannin), and makes them drinkable much sooner. Opening a bottle and allowing it to stand is one, however slow, method of aeration. Simply pouring the wine into the wineglasses and allowing the glasses to stand for a few minutes, unsipped, is another method. How often, however, do poured wineglasses sit, untouched, for minutes on end? Once the wine has been poured, most guests will begin to sip it immediately, or almost immediately.

Therefore, for young, tannic wines, decanting offers a relatively quick method for uniformily aerating the wine, as well as offering a moment of showmanship. Older wines also benefit from aeration, but the length of time that benefits the wine before it begins to oxidize varies greatly from wine to wine. That is the tricky part—how long to aerate an older wine. Once decanted, it should be tried immediately. If not quite ready, if it has a "stale" or "shut-up" or "closed" taste, allow it to stand for a few minutes. Try a sip of the wine every 15 to 20 minutes or so. After decanting, one hour is the optimum aeration time to allow for fine aged wines to "open up" unless they are relatively recent vintages when 5 to 6 hours might be preferable. Usually after an hour—sometimes even sooner—a very old wine will begin to fade.

With the improved filtration methods used by modern wine-makers, today's wines have a much better chance of not throwing sediment, or as much sediment, as wines made earlier. Nevertheless, the producing of sediment is a natural part of the aging of fine wines, especially full-bodied red wines.

As a general rule, clarets (red Bordeaux) from 1975 or earlier often benefit from decanting. Many of the ports, especially vintage port, should be decanted as well as "crusted port," which usually has quite a bit of sediment. However, not many people open a bottle of vintage port after a meal, although it is a very nice custom. Therefore, it is usually the fine, full-bodied red wines, in particular the better-quality Bordeaux, Burgundy, Barolo, and Brunello wines that are typically decanted in a restaurant setting.

Decanting the Wine

Offer the guest the choice of having the wine decanted at the cellar or at the table. Some guests may not want the rare wine to be decanted at all, preferring to pour it for themselves. The guest, however, should suggest that option; the server should always offer to decant the wine. It will be the exception rather than the rule for a guest to decline having a bottle decanted.

Why should guests have their aged wine with sediment decanted? First, guests don't want to drink the sediment—they want the clear wine. To obtain the maximum clear wine, it must be poured from the bottle into another container in one very careful slow motion so the clear liquid is poured into the decanter, leaving the sediment and a bit of wine in the original bottle.

Second, most guests who might want to decant their own bottle don't have a glass large enough to hold the contents of an entire bottle of wine and wouldn't want to use such a glass if it were available. If the guest doesn't slowly and carefully pour the wine into another container in one motion, but instead picks up the bottle, pours, replaces it, picks it up again, pours again, and so on, the sediment in the wine is being agitated and distributed throughout the wine—exactly what should not happen to a rare wine.

Remember, however, to decant or not to decant—it is the guest's choice. A server might say:

> As these wines are fragile, we offer you the choice of having your wine decanted in our wine cellar, using special apparatus to

minimize disturbing the sediment, or tableside. If we decant in our cellar, we'll bring the decanted wine, the cork, and the bottle to you after decanting. This procedure minimizes the disturbance to any sediment that might be in the bottle. Otherwise, I'll be happy to decant the wine for you tableside.

Should the customer wish to minimize the movement of the bottle, he or she might suggest that you decant at the cellar and even skip the presenting of the bottle of the rare wine to the host. This is usually acceptable only if the guest is a regular customer who has had bottles from the same case and knows the quality of that case of wine and who has recourse to management should the wine not be up to the level of prior bottles from the same case. In other words, if there is a prior existing relationship between the property management and a regular rare-wine-drinking customer, that customer might permit a bottle to be opened and decanted without first inspecting the bottle. This is not usually the case, however.

Most people will want to inspect a bottle of rare wine before the decanting process to ensure the ullage and condition of the bottle are satisfactory (cork has not been pushed out, there is no leakage, etc.). Most customers will insist that the bottle be presented, then decanted within their sight, usually in the dining room at the sidestand or on a gueridon if one is available.

When a rare wine has been ordered, obtain the bottle using the bin number on the Rare Wine List to identify where it is stored in the *Rare Wine Cabinet* or house wine cellar. If the exact location is not obvious, management should create a "rare wine bin map" that charts where each wine is stored, by both name of wine and bin number. To minimize disruptions to the temperature and humidity level, locate the precise spot where the wine is stored before opening the cabinet. Usually, only a manager, maître d', or sommelier will have a key to the rare wines, which should be kept locked at all times.

In procuring the bottle from the wine cellar or Rare Wine Cabinet, avoid disturbing sediment, especially in a red wine. Move the bottle very gently. Keep the label facing the ceiling—it should have been stored that way. If the bottle was not stored with the label facing the ceiling, remove it exactly as it was stored—if the label was facing the floor, keep it facing the floor, and so on.

Decanting with a Decanting Basket

This method evolved to ensure minimal disturbance of sediment in moving the bottle from the wine cellar to the table for consumption. The intent is to keep the bottle basically on its side, as it was

DECANTING PROCEDURE: USING
A DECANTING BASKET

Place the bottle, carefully, in the decanting basket, with the neck pointing toward the ceiling at a 30-degree angle, and the label in the same relative position as it was found in the wine cellar so as not to disturb or dislodge any of the sediment.

1. Without moving, tilting, turning, or shaking the bottle,
    (a) Trim and remove the foil capsule from the top of the wine bottle. Remove *all* of the foil capsule, not just the top portion so you will be able to watch for sediment as it approaches the neck of the bottle.
    (b) With a dampened white cloth napkin, wipe the top of the bottle clean. Especially if there is mold, which is likely, wipe off all traces of it now, before removing the cork.
    (c) Position the decanter so the wine will flow into it. Position the light source (e.g., candle or lamp) so it will shine at the shoulder and neck of the bottle. If a light, turn it on. If a candle, light the candle. (The light source can be as ordinary as a flashlight stood on end or as elaborate as a decorative "decanting" light that blends in with the restaurant decor.)
    (d) Using an "ah-so" or second choice, a corkscrew, remove the cork. An ah-so cork remover usually is gentler to the cork and may keep a fragile, crumbly cork from falling apart, whereas a traditional corkscrew screw-pull may hasten the crumbling of a cork.
    (e) Wipe the mouth of the bottle again, making sure to wipe from the inside of the bottle lip outward so any bits of cork are not inadvertently put into the bottle.
    (f) Slowly and carefully, turn the wheel of the decanting basket, tilting the bottle over the mouth of the decanter. Handle the wine very gently. Allow a steady stream of wine to flow from the bottle.
    (g) Once the wine is flowing, do not stop the wine from flowing until the end of the process or you will greatly increase the risk of mixing the sediment throughout any remaining wine in the bottle.
    (h) When about half the wine has been poured, *and*, without interrupting the flow of the wine, begin looking through the neck of the bottle, watching for sediment. The light should be behind the neck, so you are looking through the neck of the bottle.

(i) When about 80 percent of the wine is poured, a thin cloud should appear. Proceed very gently until you see the opaque stream of sediment about to enter the neck of the bottle.

(j) At this point, stop the flow. Quickly tilt the bottle upward, using the wheel.

If the decanting was performed well, only an ounce or two of wine remains in the bottle; the decanter contains the clear wine.

2. Take the nearly empty bottle, the cork, and the decanter filled with the clear wine to the individual who ordered the wine.

3. Pour a 1-ounce "tasting" sample into the host's glass. After the approval, ask, "Shall I serve the wine?" If the host says, "Yes," proceed to pour the wine in the same manner as any other wine (see Chapter Seven). Do, however, note the following exceptions:

(a) When rare wines are served, pour relatively small portions, approximately 3 ounces maximum per person. There should be 8 to 12 portions per 750 ml bottle. Rare wines are usually tasted, sip by sip, and not wildly imbibed as with "quaffing" wines.

(b) When refreshing a guest's glass, pour only a thimbleful for the slow drinkers. Assume the host will drink the greatest portion of the bottle.

4. As soon as the cork and bottle have been presented, ask, "Shall I remove the cork and bottle?" Don't be surprised if the host wants to keep either or both.

5. Offer to remove the label (with hot water) and present it atop a clean, folded napkin with the check. (This is a nice service to perform even if not requested; usually someone at the table will want the label, if not the host.)

6. Immediately after using decanting equipment, clean and return such equipment (cork pullers, decanters, candles, etc.) so another server or captain can utilize it for another bottle.

7. If anyone wishes to taste the sedimented wine, first strain it through wet, squeezed cheesecloth. If a coffee filter is used to strain the sedimented wine, a "paper taste" may be evident.

stored. When the cork and capsule are removed (without moving the bottle), the wine can then be poured without the sediment being dislodged substantially from its original position.

Decanting without a Decanting Basket

Obtain the bottle; slowly and carefully place it upright. This should be done as soon as the bottle of wine is ordered, whether it is 1 day, 2 days, 3 days, or immediately prior to consumption of the wine. Up to three days' notice will allow any sediment to fall to the bottom of the bottle, making the decanting much easier and nearly foolproof.

Another method, not highly recommended for restaurant use, is to pour the wine through cheesecloth that has been previously dampened and placed in a plastic funnel. The funnel should be resting atop the neck of a clean, empty carafe. While this method works, the showmanship is somewhat lacking. Either of the two aforementioned methods is recommended rather than the third.

DECANTING PROCEDURE: NOT USING A DECANTING BASKET

1. Without moving, tilting, turning, or shaking the bottle,
   (a) Trim and remove the foil capsule from the top of the wine bottle. Remove *all* of the foil capsule, not just the top portion so you will be able to watch for sediment as it approaches the neck of the bottle.
   (b) With a dampened white cloth napkin, wipe the top of the bottle clean. Especially if there is mold, which is likely, wipe off all traces of it now, before removing the cork.
   (c) Position the decanter to the left of the wine bottle if you are right-handed, and position the light source, in the same manner as when using a decanting basket.
   (d) Using an "ah-so" or a screwpull, remove the cork.
   (e) Wipe the mouth of the bottle again, making sure to wipe from the inside of the bottle lip outward.
   (f) Slowly and carefully, tilt the bottle over the mouth of the decanter. Handle the wine very gently. Holding the bottle very steady, allow an uninterrupted flow of wine from the bottle.
   (g) Once the wine is flowing, do not stop until you see the sediment begin to edge toward the neck of the bottle, about to be poured into the decanter. If you stop pouring

and then begin again, the resulting "wave" motion of the wine greatly increases the risk of mixing the sediment throughout any remaining wine in the bottle.

(h) When about half the wine has been poured, *and,* without interrupting the flow of the wine, begin looking very carefully through the neck/shoulder of the bottle, watching for sediment. The light should be behind the neck, so you are looking through the neck of the bottle.

(i) When about 80 percent of the wine has been poured, a thin cloud should appear. Proceed very gently until you see the opaque stream of sediment about to enter the neck of the bottle.

(j) At this point, stop the flow. Quickly tilt the bottle upward. If the decanting was performed well, only an ounce or two of wine is remaining in the bottle; the decanter contains the clear wine.

2. Take the nearly empty bottle, the cork, and the decanter filled with the clear wine to the individual who ordered the wine.

3. Pour the 1-ounce sample into the host's glass. After the approval, ask, "Shall I serve the wine?" If the host says, "Yes," proceed to pour the wine. Complete the serving of the wine following the same instructions provided for service when using a decanting basket.

❖   # RARE AND SPECIAL WINES

The following chart describes selected rare wines and shows how a typical "rare wine list" is constructed, with the types of comments, by reputed wine connoisseurs, that are usually included—the information potential buyers might like to know.

Every establishment should create its own personalized rare wine list. It could be a table tent, a portion of a larger wine list, or a separate menu noted on the main menu with an "Ask for our . . . " lead-in. It could be a decorative item regularly presented along with the food menu and wine list, in the hope that, if nothing else, it will be a conversation piece, something for guests to enjoy looking at even if they don't purchase anything.

RARE WINE TABLE

The following contains notes by Michael Broadbent, a noted wine expert employed in the Wine Department at Christie's Fine Art Auctioneers, London. Five Stars is the best rating, representing an outstanding wine.

Red Wines

Red Bordeaux

LAFITE-ROTHSCHILD

1865 Four stars. Still remarkably youthful . . . with deep vinosity and fruit . . . amazingly good on the palate.

1874 Four stars. Rich and interesting . . . medium body, firm, gentle flavor. Of all old wines this seems to benefit most from long exposure to air.

1899 Five stars. Delicate, slightly smoky; incredibly beautiful bouquet and flavour. Dryish, clean as a whistle, exquisite.

1929 Five stars. The best vintage since 1900 . . . soft, elegant wines of great finesse, and delicacy. . . . old but exquisite.

1953 Five stars. The epitome of elegance . . . an open, relaxed, fully developed bouquet, sweet cedar, fragrant . . . delicate yet generous . . .

1961 Five stars. Undoubtedly the greatest postwar vintage and one of the four best of the century. . . . subdued, restrained but rich and elegant nose, a very dry, even austere wine, full bodied, concentrated . . .

MOUTON-ROTHSCHILD

1945 Five stars. One of the greatest of all vintages, for me certainly one of the top three of this century.
Five stars. . . . fabulous and, I like to think, totally unmistakable bouquet—highly concentrated, intense black currant Cabernet-Sauvignon aroma, touch of cinnamon—flavor to match. Ripe, rich . . .

1952 Three stars. Deep, fine, flavoury with silky/leathery texture.

1955 Four stars. A beautiful, calm, dignified Pauillac bouquet, delicate, flavoury, nice balance, very attractive.

1959 Five stars. Masculine and magnificent . . . wonderfully concentrated Cabernet-Sauvignon aroma, cedar and black-currants; fairly dry, full bodied, massive yet soft, velvety, packed with flavor.

1961 Five stars. A stunning wine . . . amazing richness and ripeness of grape, concentrated Cabernet-Sauvignon bouquet, flavour; magnificant, balanced.

1966 Five stars. Plummy colour, magnificently pointed Cabernet-Sauvignon aroma; dry, stylish . . . it should be excellent.

LATOUR

1945 Five stars. Magnificently huge, opaque in appearance, a bouquet like gnarled old cedar; rich yet dry, massive, intense . . .

1949 Four stars. Deep, sweet—almost honeyed, cedary bouquet; on the palate full and rich, soft yet firm and perfectly balanced. Great depth, length of flavour and marvellous aftertaste.

1953   Five stars. Perhaps the most attractive of all the postwar vintages and, for me, a personification of claret at its most charming and elegant best.

1961   Five stars. An enormous wine . . . deep alcoholic, port-like nose; still peppery; medium dry, very full bodied, packed, beefy but beautiful balance.

1962   Four stars. A rich, soft, cedary, classic Cabernet nose; fullish, excellent flavour and balance. . . . A fine classic wine.

1966   Five stars. Magnificent colour; bouquet and flavour packed and closed in, but an enormous, rich wine. Dry, yet velvet-lined.

PICHON-BARON

1955   Three stars. A good but under-appreciated vintage . . . has quietly blossomed in recent years.

1961   Five stars. Undoubtedly the greatest postwar vintage and one of the four best of the century.

MARGAUX

1945   Five stars. Intensely rich and fragrant bouquet, classic cedar-pencil/ cigar-box character . . . velvet textured and magnificent . . .

1947   Four stars. Still magnificently deep, a big ripe hot-vintage colour; bouquet to match, very mature but no signs of decay . . . velvety, towering yet delicate, rich yet dry.

1952   Four stars. A fairly massive wine, velvety with, seemingly, a whole cask of bouquet and flavour waiting to burst out . . .

1953   Five stars. Margaux at its best . . . magnificent bouquet; rich, waxy, elegant, soft and silky, excellent balance . . . long fragrant aftertaste.

1959   Four stars. Rich, cedary scented bouquet . . . fairly full bodied, rich, velvety, certainly no lack of acidity.

1961   Five stars. Intense fragrance . . . the bouquet is extraordinary—rich, singed creosote, intense, lovely; rich, silky, elegant, long flavour, very dry finish.

1962   Four stars. Somewhat delicate wine . . . fragrant, smoky, very fine bouquet and flavour. Richness and delicacy.

1966   Five stars. Rich, fruity, elegant bouquet and flavour, some plumpness but slim, elegant yet with fine structure.

LÉOVILLE-LAS-CASES

1962   Four stars. Almost opaque colour; fine, iron-tinged, sweet bouquet and flavour. Most attractive.

GRUAUD-LAROSE

1959   Four stars. Fine, deep, mature colour, heavy bead (legs); rich, meaty . . . very flavoury.

COS D'ESTOURNEL

1959   Four stars. Beautifully made, well-balanced. A full, firm, beefy wine.

1961   Four stars. Magnificent depth of colour, bouquet and flavour. Very tannic.

1966   Four stars. Deep, tannic.

HAUT BRION

1934   Three stars. Undoubtedly the best vintage of the 1930s.

1945   Five stars. Absolute perfection . . . fine, rich, fragrant and complex bouquet; slight sweetness of wine made from fully ripe grapes, magnificent, chewy, chunky yet smooth. Magnificent aftertaste.

1949   Four stars. Elegant and rich . . . deeper that Latour; sweet, rich, velvety nose; very rich, complex earthy wine . . . sound, characterful.

1953   Five stars. Lovely rich Graves-earthy aroma and flavour; great depth, richness, vinosity.

1955   Four stars. Soft gentle bouquet; Graves earthy, slightly charred flavour with typical Haut Brion elegance.

1959   Five stars. Deep, magnificent tobacco-like bouquet; excellent earthy heavyweight with a dry tannic finish.

1961   Five stars. Characteristic hot, earthy/pebbly bouquet . . . ripe, lovely texture . . . understated.

CHEVAL BLANC

1947   Five stars. A complacent, abundantly confident bouquet, calm, rich distinguished . . . slightly sweet, plump, almost fat, ripe, incredibly rich . . . a magnificent wine, almost port-like.

1959   Four stars. Sweet and fruity bouquet, ripe, full, rich, soft. Dry finish. Fully developed.

1961   Four stars. Mulberry aroma, rich, port-like (reminiscent of the '47); distinct sweetness of fully ripe grapes, soft, open . . . gentle but firm finish.

1966   Four stars. An excellent vintage. Stylish, elegant, well-balanced. Lean rather than plump, though with good firm flesh . . . Bordeaux at its most elegant.

PETRUS

1961   Five stars. Extremely deep coloured, magnificent bouquet, amazing fruit; a lovely, rich, velvety wine.

Red Burgundy

ROMANÉE-CONTI

1952   Five stars. Very rich, complex, excellent pinot bouquet; rich, fine flavour and balance. Real Romanee quality. Holding well.

1959   Five stars. Nose slightly aloof and withdrawn, excellent but still hard; a huge, dry tannic ald alcoholic wine, almost Latour massiveness.

1966   Five stars. Very rich bouquet; ripe and rich on the palate, firm, even a little austere, with a dry finish. Clearly years of development ahead.

LA TACHE

1970   Three stars. Light and sugary aroma; slightly sweet, soft yet some acidity. But flavoury as always.

GRANDS ECHEZEAUX LEBEGUE-BICHET

1959   Five stars. A great vintage and, in my opinion, one of the last classic heavyweights made in Burgundy.

White Wines

CHATEAU D'YQUEM

1784   According to the *Wine Spectator*, a bottle of 1784 Yquem was believed to have been ordered by Thomas Jefferson. It recently ('89) sold for $56,000—the highest price ever paid for a white wine.

1811   Bottle of excellent condition wine sold for $30,500 at Christie's per the *Wine Spectator*. This bottle was from the "comet vintage," named for the Napoleon comet. Considered to be the finest d'Yquem of the 19th century. Recorked at the chateau in 1987; label and capsule in excellent condition.

1967   Five stars. Pure amber gold . . . honeyed, overripe, botrytis grape smell; holding its sugar, fairly full-bodied, rich pronounced flavour.

*Reprinted with permission.

T w e l v e

Organic Wines

Organic wine? Isn't all wine organic? The answer to that question is yes, to some degree. It's the question of degree that grape growers, winemakers, distributors and consumers discuss and argue about incessantly. While the grapes are, themselves, organic, is the growing of the grapes done in an organic manner? Are chemical pesticides and fungicides used? Are phosphate fertilizers used? Are the grapes treated with sulfur dioxide? Does the use of any chemicals in growing the grapes used to make the wine result in nonorganic wine? If so, which chemicals? To what degree?

❖ DEFINING ORGANIC WINE

In the processing of the grapes, what constitutes organic? What materials are used for filtering the wine? Clarifying the wine? Do bleached corks qualify as "organic"? Do tin and lead bottle capsules contaminate or endanger the health of the consumer? Should those be listed as nonorganic substances? What about stainless steel rather than oak tanks? The questions, and the debates, continue.

Many view the issue as having two major components:

1. How are the grapes grown, organically or with the use of pesticides, fungicides, and chemical fertilizers?
2. How are the grapes processed into wine once they are harvested?

Many grape growers and winemakers who grow their grapes "organically," which is 90 percent of the making of the wine, do not use chemical pesticides, fungicides, and chemical fertilizers. They label their wines as made from "organically grown grapes."

For the majority of wine consumers, "organically grown grapes" that are made into wine using the same traditional, sound winemaking processes used in making other fine wines, suffice for the wine to be labeled "organic" wine and for them to feel that they are drinking such wine.

❖ GROWING GRAPES
ORGANICALLY

Producing organically grown grapes is a significant undertaking. Even if a grape grower does not apply chemical pesticides, fungicides, and chemical fertilizers, water from other sources (including runoff from nearby mountains, streams, and even rain) may contain chemical residues that can affect the grapes.

In California, grape growers and others concerned with organic farming formed a group called the *California Certified Organic Farmers* (CCOF). CCOF defines organically grown grapes as, among other criteria, those grown without the use of chemical additives, including chemical pesticides, fungicides, and chemical fertilizers, and which, upon harvesting, have no detectable chemical residues. Furthermore, the grape growers belonging to CCOF agree to inspections from CCOF members and agree to farm their land so as to maintain long-term fertility of the soil, enabling generation after generation of crop production from the same land.

Modern organic farmers and grape growers are using innovative farming methods that include dry farming the vines and plowing and hoeing in two directions. Using bugs to assist rather than trying to eradicate them, as has been done in the past, is another innovation. *Konrad Estate* uses insects such as ladybugs and predatory wasps instead of pesticides. They use chemical-free, organic compost rather than chemical weed killers and organic mulch rather than chemical fertilizers. *Redwood Valley Vineyard* uses cover crops watered with drip-irrigation and blackberry bushes to protect the grape crops. The grape leafhopper, a pest that strips the foliage from grape vines, can be controlled by a wasp, anagrus, that attacks the leafhopper.

Because a number of chemical sprays, fungicides, pesticides, and herbicides are potentially toxic, it is important for consumers to recognize that whether or not grapes are grown organically is a significant issue. California's Department of Food and Agriculture lists insecticides, pesticides, fumigants, herbicides, and other chemicals that are toxic or highly poisonous. Seventeen chemicals commonly used in the growing of grapes are considered toxic. Organic farming, or "what's been put into the soil," becomes the central issue to the production of organic wine, keeping the grapes safe and healthy for everyone.

❖ ACCEPTABLE LEVELS OF SULFITE IN WINE

To a very small number of people hypersensitive to sulfites, and to the U.S. Department of the Treasury's Bureau of Alcohol, Tobacco and Firearms (ATF), however, a more stringent level of disclosure regarding chemicals produced during winemaking and added to wine (e.g., sulfites), particularly sulfur dioxide, is necessary. The ATF level below which a wine label need *not* state that the wine contains sulfites is *less than* 10 parts of sulfur dioxide per million parts of wine. If more than 10 parts per million are contained in the wine, the label must state "contains sulfites."

A level of sulfites in the range of 100 parts of sulfur dioxide per million parts or less is generally accepted as "reasonable" by winemakers and producers within the industry. This level of sulfur dioxide does not affect the taste of the wine, but enhances the wine by enabling it to age well, maintaining its complexities and body. Many winemakers contend that without the judicious use of sulfur dioxide it's extremely difficult to create quality wines that are consistent, stable, and will hold up over time.

Winemakers today utilize refined winemaking techniques and sophisticated filtration processes enabling them to use minimal amounts of sulfur dioxide in the processing of wines. Amounts of sulfur dioxide used in the making of wine today are generally two-thirds less than used 50 years ago. While winemakers are very competitive, each trying to make the "best" wine, they cooperate in sharing the best and newest viticultural methods to produce better-tasting, longer-lasting wines. The majority of winemakers are concerned with health and ecological issues as well as quality issues. Today's winemakers want to produce the best grapes they can grow using the least amount of additives, natural or chemical, that they can. It's only by starting with the highest quality grapes that they can produce the best possible wine.

❖ REGULATION OF WINE PRODUCTION

California, under the *Organic Food Act of 1979*, established official guidelines for "organic" farming. This law, however, did not specifically address the complexities of winemaking, nor did it broach the "What constitutes organic versus nonorganic?" question. The

California Organic Food Act of 1990 revised the 1979 law but did little to clarify it. Currently, Title XXI of the *Food, Agriculture, Conservation and Trade Act of 1990,* better known as the "Organic Foods Production Act of 1990," Public Law 101-624, Sections 2101–2123, is the definitive federal regulation regarding the standards for and definition of organic products, including organic wine. However, it is not yet definitive on many issues.

Taking a broad perspective, the Organic Foods Production Act of 1990 was passed with the intent to establish federal standards regarding organic foods and food production: What can and cannot be used in organic farming? If something can be used, how much of it? How should it be applied? and so on.

While the Act has been passed, it is not yet clear regarding specific compounds, chemicals, processes, or additives that may or may not be used, nor on enforcement issues. Enforcement of organic standards, once set, will be under the purview of the United States Department of Agriculture (USDA).

Currently, the *National Organic Standards Board* (NOSB), having been created by the Act, is gathering information and obtaining input regarding standards and detail-level information concerning organic standards. Once the Board reviews and synthesizes its information, recommendations will be made, passed to the USDA for further review and input from the public and interested parties, and then will be enacted and enforced. Review of and enactment of specific recommendations is projected to be within the 1994–1995 time frame.

Regulations in Foreign Countries

Europe has been producing and regulating organic wines for some time. In 1982, Veronique Raskin, who was one of the first to publicize and distribute organic wines in the United States on a broad scale, began importing her grandfather's organic wines from France. Her grandfather and other winemakers with small, family estates in Europe had become concerned about the use of chemicals. Some, including Patrick Boudon's father, a maker of French organic wines from their family estate, *Domaine Ste.-Anne,* noticed that the fungicides they were using seemed to be killing the natural yeasts on the grapes. Without the natural yeasts, the natural fermentation, which is necessary for the production of wine, was slowed and other yeasts had to be added to compensate.

France, over 30 years ago, established organic wine standards. Organic winemakers are supervised by one of three major "certification" organizations: *Union Nationale Interprofessionelle de l'Argobiologie, Nature & Progrès,* or *Terre & Vie.* These three organizations are then supervised by the French Ministry of Agriculture.

In France, it is recognized that certain minerals and trace elements in the soil and other natural ingredients and additives have traditionally been used in the making of wine for centuries. Many of these ingredients and processes combine to define the characteristics of the soil, becoming part of the microclimate novel to each vineyard, each château, bottler, or winemaker. Natural materials that have been used in the past are typically permitted today in the processing of grapes into organic wine; for instance, manure compost, mined sulfur, copper sulfate, and lime are permitted in Bordeaux. However, the use of chemicals that might be taken into the plant's root system or into the plant through the leaves, such as synthetic herbicides, fungicides, pesticides, and fertilizers, are not permitted. While synthetic chemicals are not allowed to be used on grapes while they are growing, during the production of organic French wine, the use of sulfur dioxide *is* permitted to prevent oxidation, browning, and bacterial spoilage. In the past, however, this sulfur dioxide had to be derived from mined sulfur only. It could not be sulfur dioxide recovered from spent coal and oil combustion sources (note that these sources are a cheaper and purer source of sulfur dioxide). Today, both mined sulfur and sulfur gas bubbled through water to produce sulfur dioxide are used in the production of organic French wines.

While the use of sulfur dioxide is permitted in France for the making of organic wines, the use of sulfur dioxide seems to be one of the major stumbling blocks for those who want to declare whether or not a wine has been organically produced in the United States.

Arguments for Using Sulfur Dioxide

Sulfur dioxide is deemed essential by many winemakers for making fine wines that have full flavor and can age well. It prolongs the useful life of the grapes, preventing spoilage, oxidation, and browning, especially during the harvesting and initial stages of winemaking. Many experts, including Cornelius S. Ough, a professor of enology at the University of California at Davis (said to be the world's premier winemaking, wine-studying, and wine-knowledgeable campus), have realized that during the winemaking process, sulfur dioxide is produced naturally as a by-product of the fermentation process. Other studies have reached the same conclusion, even indicating that it is possible for the fermentation process alone to produce sulfur dioxide exceeding the limits set by the ATF.

Consider the case of six California wineries that used organic methods both in growing grapes *and* in processing the wine—these wineries used organically grown grapes *and* added no sulfur dioxide. In one year, only three of the wines produced at these six wineries met the level set by the ATF. The other wines produced by these wineries had higher levels of sulfur dioxide than 10 parts per million and so were required to carry sulfite warnings on their labels.

During the late 1989s and the early 1990s, the *Organic Grapes into Wine Alliance* (OGWA), an organization of people interested in promoting and establishing standards for organic wine, was organized by Brian Fitzpatrick of the *Fitzpatrick Winery.* The OGWA members support the use of organically grown grapes, as well as the judicious use of sulfur dioxide, in making their organic wines. OGWA supports standards allowing up to 100 parts of sulfur dioxide per million in wine that is labeled "organic." (Currently the legal limit is 350 parts per million.)

Many OGWA members believe that by allowing a judicious use of sulfur dioxide, which is produced naturally during fermentation, the use of pesticides and other chemical additives can eventually be eliminated. Not permitting the addition of sulfur dioxide in the production of organic wines leaves the door wide open for wine producers to use any number of other, not necessarily organic, additives to accomplish the same processing result.

Most organic grape growers and winemakers are lobbying for permission to use sulfur dioxide in making organic wines while maintaining a separate subset of "no-sulfites-added-organic" wines for those with a hypersensitivity to sulfites.

OGWA is lobbying for the use of sulfur gas bubbled through water as the source of sulfur dioxide in the production of wine. Bubbling sulfur gas through water is currently the preferred source of sulfur dioxide for French organic wines. OGWA members and other organic grape growers are opposed to the use of potassium metabisulfite as the source of the sulfur dioxide.

The Advantages of International Standards

The implications are exciting if the United States adopts and implements standards for organic wine that compare with the standards that have been enforced in France for over 30 years. We'll have opened doors for marketing wines internationally, throughout the Common Market, and vice versa. The choices to consumers will be increased dramatically.

Many wineries are experimenting with growing grapes organically and have devoted part, or all, of their land for this purpose. Wineries and wine producers experimenting with organically grown grapes include Fetzer (Chardonnay and red organic table wines), Las Montanas Winery, Buena Vista Winery, Robert Mondavi Winery, Fitzpatrick Winery, Hallcrest Vineyards, Frey Vineyards, Konrad Estate. In addition, well over 20 organic wines are being imported from France, Germany, and Italy. Organic wines are being made in Australia and New Zealand as well. Winemakers there are awaiting final resolution of the Organic Foods Production Act of 1990 specifications and standards as recommended by the NOSB. Once those standards have been set, other countries can meet our standards and begin strongly marketing their organic wines in the United States.

If international standards can be established, winemakers throughout the world would have a set of uniform wine production standards. This would make it easier for consumers in any country to know what they are purchasing when they order a bottle of wine. Typical wine lists in American restaurants today have wines from several countries, and this is expected to continue. As wine lists expand to include the subcategory of organic wines, it would be helpful for consumers to have an international set of standards defining an organic wine. Having a clear understanding of what organic wine means is essential for tomorrow's leading-edge restaurateurs and other wine purchasers, including guests. This knowledge is needed not only for American wines, but also for organic wines produced in France, Italy, Germany, Australia, New Zealand, Chile, or anywhere else in the world. Customers want to know what they are ordering.

MARKETING ORGANIC WINE AS A DIFFERENCE MAKER

Restaurants specializing in organic foods and beverages are constantly seeking new and additional organic wines to add to their beverage lists. One such gourmet restaurant, *Nosmo King* in New York City, offers not only organic varietal-grape juices but organic and chemical-free bottled wines and natural sodas. And their clientele is growing. Perhaps that growing clientele base has been aided by their unique marketing techniques, including wine tastings of organic wines accompanied by press receptions. What a great way to get publicity for a restaurant, obtain word-of-mouth advertising, and give the guests a reason to return!

Tastings of no-added-sulfite wines have led some to conclude that these wines are light to medium in body and immediately drinkable, not meant to be aged. Some have alluded to the flavor of no-sulfite-added wines as "unusually fresh and light." The wines have even been referred to as having a "doughy" flavor.

Many American producers of organic wines are labeling them "wines made from organically grown grapes." This enables them to use the finest available organically grown grapes with no detectable chemical residues and, in California, grapes that meet the standards set by the CCOF for environmentally safe farming. These vintners then may (or may not) judiciously use sulfur dioxide in producing the wine.

Those who market organic wines affirm that this market is definitely growing. In 1989, California produced and sold approximately 50,000 cases of organic wines; that number doubled in one year. Wines being produced from organically grown grapes include several popular wines, including Zinfandel, Cabernet Sauvignon, Chardonnay, Syrah, Merlot, Champagne, Riesling, and Colombard varietals as well as table wines. Many producers of quality wines are planning to convert some of their vineyards to organic agriculture in the production of their grapes. Much of the Fetzer grape production is already organic, especially around its Hopland Valley Oaks Food and Wine Center. Fitzpatrick Winery has been producing organically grown grapes and wines for 14 years. Konrad Estate has been growing grapes organically for 4 years.

The market for organic wines is growing, consumer awareness of the dangers of chemicals and tainted foods is growing, and the need for ecologically sound methods of producing our foods is growing. These winemakers are meeting that challenge. Consider adding wines made from organically grown grapes on your wine list; consider organic wine tastings; market an organic wine as the "Wine of the Month."

A PRIMER ON WINE SERVICE

T h i r t e e n

Wine: What Is It?

How is wine made? What are "good" wines? What is champagne? Because beer and wine—whether alcoholic or dealcoholized, natural (organic) or made with chemical additives—have significant sales potential, restaurateurs need to be knowledgeable about these beverage alternatives so they can market them to guests. The service of beverages, including wine, was discussed earlier; in this chapter, we'll concentrate on the characteristics of wine, the production of wines, and various distinctions among wines.

Most of the wine purchased and consumed in the United States is made from grapes, although some "wines" are made from other fruits (strawberry wine) or flowers (dandelion wine). Only grape wine should be referred to simply as "wine"; wines from other fruits must be qualified by the fruit's name ("cherry wine").

There are many grape varieties suitable for making wine. Some grapes produce better wine than others, and some grape varieties are better for eating than for wine making. Some grapes grow best in cool climates, some grow best in hot climates. Many factors influence the final taste of the wine:

♦ The grape variety.
♦ Where the grapes were grown for that individual bottle—
    Alkaline soil? Gravelly or sandy soil?
    Slope of the vineyard? Hill or valley?
♦ Weather that season—
    Number of days of sun?
    Amount of rainfall? Temperatures?
♦ Harvesting and winemaking procedures—
    Was the juice kept in contact with the skin?
    Was the wine aged in oak barrels? Stainless steel tanks? Both?
    If in oak, was it young oak? For how long?

These and other factors are discussed in this chapter. Also included are factors in the selection and enjoyment of wine. Many

235

wines are sold as "varietals," that is, based on the variety of grape primarily used to make that wine. Some of the more popular wine varietals are Chardonnay, Cabernet Sauvignon, Riesling. The chart shows the principal grape varieties used in the production of wine.

❖ VARIETIES OF GRAPES

The quality of the wine and 90 percent of its character depend on the variety, selection, planting, and cultivation of the grape from which it is made. While most people don't have the opportunity or desire to make their own wine, thereby learning about various grapes' characteristics, knowledge about winemaking and the more common grapes can provide clues to determine a wine's probable taste when purchasing wines that have not been tasted previously.

Different wines made from the same grape variety will not taste identical but should be in the same "taste family." For example, a person with a good "taste memory" can remember the taste of Pinot Noir grapes (used in Burgundies in France and wines in many other countries) and will be able to identify the "Pinot Noir" grape taste in wines, even though each wine tastes slightly different. If a label states that Pinot Noir grapes were used in a wine, a person with a taste memory of Pinot Noir will have an idea of the taste of that wine.

The color of the grape is not necessarily the color of the wine. Both white and rosé wines may be made from red or black grapes. The taste of the wine is usually a better indicator of the grape variety than the color.

❖ PRINCIPAL ELEMENTS OF WINE PRODUCTION

Cultivating Grapes

Making great wine is very much dependent on the soil and microclimate of each grape-growing region and, in the case of the great limited-production wines, the exact location of the vineyard, in addition to the cultivation methods of the winegrower.

Cultivation of the vine bearing the grapes influences the yield of wine and the quality of the product—the greater the yield per acre, the less flavor in each grape, and a "lesser" or more diluted

PRINCIPAL GRAPE VARIETIES FOR WINE

Grape Name	Grape Color	Wine Name	Wine Color	Country
Aligoté	W	Burgundy	W	Fr, US
Alvarelhao	B	Port		Po
Barber	B	Barbera	R	It, US
Bastardo	B	Port		Po, Fr
Bouchet		Saint-Émilion	R	Fr
Boal	W	Madeira		Po
Cabernet-Sauvignon	B	Saint-Émilion	R	Fr
		Médoc		
		Cabernet-Sauv	R	US, It, Au
Canaiolo	B	Chianti	R	It
Chardonnay	W	Champagne	W	Fr, US
		Burgundy	W	US
Chenin Blanc	W	Vouvray	W	Fr
		Saumur		US
Furmint	W	Tokay	W	Hu
Gamay	B	Beaujolais	R	Fr
Gewürtztraminer	W	"Spicy" Traminer	W	Fr, Gr, US
Grenache	B	Côtes-du-Rhône	R	Fr
		Tavel Rosé	Rz	Fr
Grignolino	B	Varietal	R	It, US
Insolia	W	Marsala		It
Malbec	B	Bordeaux	R	Fr
Malvasia	W	Malmsey	W	It
		Madeira		Po
Merlot	B	Bordeaux	R	Fr
Mission	B	Port		US
Muscadelle	W	Sauterne	W	Fr
Muscat	B,W	Asti Spumante	R	It, Fr
		Muscatel	W	US
Nebbiolo	B	Barolo	R	It
		Barbaresco	R	It
Pinot Blanc	W	Champagne	W	Fr, US
Pinot Noir	B	Chablis	W	Fr, US
Rabigato	W	Port	W	Po
Riesling	W	Rhine	W	Fr
		Moselle		Gr
		Alsatian		Sw, Au, US
Sangiovese	R	Chianti	R	It
Sauvignon Blanc	W	Graves	W	Fr
		Sauterne	W	US
Sémillon	W	Sauterne	W	Fr, US
Sercial	W	Madeira		Po
Traminer	W	Varietal	W	Fr, Gr
Trebbiano	W	Chianti	W	It, US
Zinfandel	B	Zinfandel	R	Ca

Key: Fr France   Sw Switzerland   It Italy
      US United States   Gr Germany   Au Australia
      Po Portugal

wine. With great wines, the growers limit the yield per acre, usually by pruning the vines. In some places, the maximum yield per acre is dictated by law.

The age of the vine also influences the quality of the grapes. Very young vines do not have a well-developed root system reaching down into the soil for the nutrients and minerals that give the grapes their wonderful, full-bodied taste. Consequently, grapes from very young vines (under 4 years old generally, up to 10 years in some instances) will not be as flavorful as grapes from older vines—even if from the same vineyard.

The wine's essential character is attributable to the marriage of vine and soil, complemented by the weather—rain, sun, and wind. Replenishment of nutrients to the soil is essential, which explains why portions of some vineyards lie fallow some years, being fertilized and nourished, then are planted with young vines the following year. Then begins the 5-year (or more) wait until the vine produces flavorful grapes.

Fermenting and Aging the Wine

Vinification, or winemaking, is the next phase of wine production that has a great influence on the taste of the wine. It is impossible to create a great wine from mediocre grapes, but it is easy to do the opposite, or even to ruin a potentially great wine if the grapes are not handled well or production methods are sloppy.

Wine is made from the juice of freshly picked grapes, either of a single variety or of several. The grapes are crushed, and almost immediately the first of two fermentations begins. Yeasts, which are single-celled microscopic plants that are present in the air and on the grapes, begin to convert the sugar in the grape juice to alcohol, releasing carbon dioxide. This fermentation, in casks or vats, may last 2 weeks. The grape juice is referred to as "must."

By this point, the vintner has engineered a great deal of the wine's quality. If the grapes picked were ripe or overripe, they are high in sugar and low in acid, and the grape juice (must) will also be high in sugar. The resulting wine will have a high alcohol content and perhaps even residual sugar. If the skins have been left with the grapes, the wine will take on the astringent quality of the tannins contained in the skins as well as their color. The color is released through contact of the fermenting grape juice with the grape skins.

Throughout the fermentation process, the temperature of the wine is controlled, accelerating or decelerating the action of the yeast. The fermentation process continues until all the sugar in the grape juice, or must, has been converted to alcohol, or, until the alcoholic content is so high (usually about 14% by volume) that the yeasts no longer function. A solution with an alcohol concentration above 14 percent will kill the yeast. The fermentation may also be stopped by adding chemicals, by regulating the temperature, or by adding alcohol (in the form of brandy). Stopping the fermentation process before all the natural sugar in the grape juice has been converted to alcohol has the effect of leaving some sugar in the finished wine.

Later, a second fermentation by bacteria takes place in casks or vats, except for Champagne and other sparkling wines, which are yeast fermented for the second time in the bottle.

After the second fermentation, the wine is placed in storage casks or vats to age, or mature. The extent of aging is determined by the particular wine and the particular qualities of the grapes of that season. It may be a matter of a few weeks for some wines, or many years for others. At this point, the wine is left alone and checked occasionally to judge when it is ready for bottling. The most radical and distinctive changes occur during this storage period—not, as many individuals imagine, during further maturation in the bottle. Bottling largely stops the aging process.

When the wine in the cask is ready for bottling, having developed flavor, brilliance, and clarity, it is filtered (either naturally with egg whites, or with filter paper, or mechanically, or chemically) and bottled.

Prolonging the aging does not necessarily result in a continued improvement of the wine. Each wine has a definite peak noted by a balance in acidity and fruit flavors, after which the wine declines—usually the fruit flavor decreases while the acidity goes up. Likewise, there is no virtue in age per se, either in the cask or in the bottle: Wine X may be "excellent" in 3 years, and wine Y may be the same in 40 years. Age as a standard of evaluation has meaning only in comparing bottles from the same vineyard, of the same production batch, of the same production year, but tasted in different years.

Once bottled, most wines are ready for drinking. However, some require further aging, especially red wines from "hearty" grapes (e.g., Cabernet Sauvignon or Pinot Noir). Depending on the wine and how it is handled and stored, a year, a decade, or a half century may be the "right" amount of time to age the wine in the bottle.

How to tell if the wine is ready to drink? First, establish the earliest it is acceptable to drink it, based on the type of wine and the opinion of the growers and other wine experts. Historically, certain wines are drunk "young"; others require long maturation.

When this threshold has been established, consider the storage location. Is it cool (55°F–62°F), dark, quiet, and with a relative humidity of 55 percent? If not, adjust—the wine will be ready to drink sooner if the storage place was warm, exposed to light, high in humidity or vibrations, and so on.

Then taste the wine. Determine the wine's peak by tasting bottles periodically, such as every 6 months or every year and keeping "tasting notes" on the acidity, bouquet, and fruitiness, or by following the taste experiments of an expert, either the purveyor or a wine writer.

When there seems to be a good balance of acidity to fruitiness and the tannins add body and depth but do not overpower the wine, it's time to drink the wine. Don't wait until it is "over the hill."

❖   TYPES OF WINE

Distinguishing among wines can be confusing. It can be helpful to classify them according to certain characteristics:

◆ Type of grape.
◆ Country of origin.
◆ Microclimate (exact location, type of soil, etc.).
◆ Sparkling or still.

Classification by country of origin is helpful for generalities, combined with the grapes used or a well-known wine name or bottler or shipper. People often know a few major types of wines from each wine-producing country and can use them as a starting point, "Oh, is this similar to . . . ?"

Four different systems of wine classification have been developed over the years. They can be further refined by distinctions apparent from the labels. These systems use the following classifications:

1. Use.
2. Production method.
3. Color.
4. Recognized "quality standard."

The next four sections describe each of these systems.

Use

Aperitif Wines

Specifically created to be drunk before a meal or as a "cocktail," these beverages include spiced wines, such as vermouth, and various brand-name preparations, such as Dubonnet, Campari, Punt e Mes, and Lillet. Sherries and similar wines, which may be drunk during a meal but are not likely to be, are also included in this category.

Table Wines

These wines are enjoyed with food. A table wine can be inexpensive or expensive, red or white (or somewhere in between), still or sparkling. This category is certainly the largest.

Dessert Wines

Some wines have sufficiently high sugar content to be drunk with or after a dessert. In general, as aperitifs they would blunt the appetite; with a main dish they would taste strange. Dessert wines range from Sauterne to Port to Trockenbeerenauslese—a very wide range.

Refreshment Wines

"Fun" wines or "quaffing" wines are created to be drunk unceremoniously, often without food accompaniment, without even the anticipation of a meal. They are drunk instead of water, milk, or punch, beer, fruit juice, or cola. Many new wines are sold under upbeat brand names and are meant to be chilled ice cold or drunk on ice as refreshment wines.

Production Method

Natural Wines

This method involves the simple fermentation of grape juice, without the addition of alcohol or sugar (beyond the very small amount of sugar allowed by law to supplement the natural sugar of some wines in certain years). Natural wines are usually dry wines.

Fortified Wines

These wines have had alcohol added to them, with two results:

1. A sweetness due to unfermented sugar present at the time the added alcohol inhibited further activity by the yeast.
2. An alcoholic content (or "proof") beyond that of natural wine. A fortified wine may have 20 percent alcohol, in contrast to a natural table wine, which is usually between 7 and 14 percent alcohol by volume.

Sparkling Wines

Carbonation, caused by the presence of carbon dioxide in a wine, is created by a change in production method. In the case of Champagne, it is the corking of the bottles for the second fermentation, preventing the carbon dioxide released during the fermentation to escape. Or, if bulk method, it is the addition of carbon dioxide bubbles to the wine in an automated, mechanical way, forcing the wine to become "supercharged."

Made Wines

These are blends of wines, or of wines and other natural or artificial products to meet a brand standard. These include wine cocktails, the best Champagnes (the most notable examples), and South African brand-name drinks that combine cola syrup, brandy, and red wine (the most dramatic).

Color

Color can be a very meaningful distinction between wines, or it can be a distinction without value. It is of value to a person who knows a particular wine's usual hue and its intensity. If someone knows the shade or shades and the intensity of red that typify a red Burgundy, a bottle that does not fit within the normal spectrum will alert the person that:

♦ It may not be a red Burgundy.
♦ It may be an improperly stored red Burgundy that has gone bad.
♦ It may be a very poorly made red Burgundy, in which case it should not be carrying that label, but should be sold as "red table wine."

The color distinction is useful when it communicates other distinctions or leads to distinguishing questions. For example, white Graves is a famous and often expensive wine for reasons that have nothing to do with color; red Graves is less well known and is a totally different type of wine. Color would be an important distinction were a merchant offering cases of Graves. Likewise, white Port might be distinguished from Ruby Port, or white Chianti from Chianti.

Recognized "Quality Standard"

Major world-renowned producers, (e.g., France, Italy, and Germany) award designations of relative merit to wines that meet certain standards of quality and genuineness.

Although there is tremendous variation within these categories, the words *Denominazione di Origine Controllata* (on an Italian wine) or *Appellation Contrôlèe* (on a French wine) mean that the wine is genuine, that it is produced in the region the label indicates in the year indicated, and so on. In other words, the wine has a certain level of quality, although precisely what that level of quality is is not always determinable.

The Germans are much more explicit; for example, the *Qualitatswein mit Prädikat* are quality wines with "special attributes." Then those attributes are described using several levels of special qualities (and sweetness). A wine with those labels is telling the purchaser exactly what to expect when the bottle is opened.

❖ GUIDE TO PURCHASING WINE

The purchase of wine involves many considerations such as taste and price as well as the laws controlling the sale of alcoholic beverages, including wines. Various federal, state, and local regulations govern the importation and sale of wine. Some restaurateurs cannot sell wine; others cannot sell it by the glass. Others can buy it in the bottle but not in the cask.

The price may be fixed, supported by a minimum markup, or not fixed at all. The extent and type of taxation is also an important price factor; taxes, and their forms, vary throughout the country.

Wine Packaging

The largest bottle that the individual consumer or restaurateur is likely to encounter is the magnum, equivalent to two average bottles (51 ounces or 1.5 liters), and the smallest is the nip or baby champagne (6 to 6½ ounces). There are larger bottles, but they are seldom-seen curiosities, used mainly for show at a celebration, wedding, or other special event.

Most commercial bottles fall between these two extremes. Exact contents must be stated on the label, the usual bottle being 750 ml or 1 liter.

There is some relation between the size of the bottle and the quality of the wine it contains. The standard 750 ml bottle with a cork is the most common for quality wines. If the same wine is put into a small bottle it will mature more quickly; in a magnum, more slowly. Wines in bottles with screw-cap closures, or flat mushroom corks, do not mature in the bottle.

The shape of the bottle may be an indication of the wine it contains. Burgundy and Bordeaux wines, for example, are packaged in characteristic bottles.

Wine can also be purchased in other containers. Cardboard boxes with plastic inserts that deflate as the wine is used are becoming increasingly popular as there is very little wine-to-air contact as the container empties, keeping the wine fresher and better tasting. Cans, similar in all respects to soft-drink cans, are also in use, as are plastic bottles and flexible heavy plastic sacks.

The type of bottling can be an important purchasing consideration. The cost of bottling, shipping, and handling a bottle of inexpensive wine is the same as the cost for an expensive wine. As the price of the wine increases, the fixed costs remain the same, and the percentage for packaging decreases.

Essentially, this means:

1. Inexpensive wines are better bargains in large containers.
2. Inexpensive wines in bottles may not be bargains at all, since a relatively small additional sum can buy a much better wine.

Bottles commonly used for wine packaging are listed in the accompanying chart.

Buying by the Label

The wine label usually provides significant information, some of which is important for basing a purchase decision; these factors

COMMON WINE BOTTLE PACKAGING		
Name of Bottle	*Capacity (ounces)*	*Common Uses*
Nip, baby, split	6–6½	Champagne, sparkling wines
Split	8	Dinner wines, Chianti
Half bottle	12–13	Dinner wines, Champagne
Half bottle (tenth)	12.8	Refreshment wines, dinner wines, aperitifs (Sherry, Port)
Pint	16	Refreshment wines, Sherry, Port, Muscatel
Imperial pint	18	Champagne
Bottles	24–26	Wines of all types, usually 750 ml.
Fifth (⁴⁄₅ quart)	25.4	Wines of all types
Bottle	30	Chianti, vermouth
Quart	32	Wines of all types
Magnum	52	Dinner wines, Champagne (used occasionally)
Half gallon	64	Wines of all types
Gallon	128	Wines of all types

must be understood and interpreted to be useful. Wine labels, however informative, leave an essential question unanswered—"Is the wine any good?"

The bottle itself offers clues—the color of the wine, the wine level within the bottle (ullage), the size of the bottle, even the number of ounces in the container. On the label, there may be 20 or so items that can be used to evaluate the wine before buying it and tasting it.

In large part, for an unknown wine, the label helps only to formulate some important questions to be answered, either by research or by experimentation. Otherwise, the label offers a series of clues to the quality and the character of the wine; some of them are of sufficient weight to influence a decision to buy the wine. As many as 23 types of information may appear on a bottle:

1. Name of the wine.
2. Country of origin.
3. Color.
4. Year grapes were crushed.
5. Region where grapes were grown and crushed.
6. Kind of wine.

7. Percentage of alcohol.
8. Liquid ounces in bottle.
9. Official and governmental guarantees.
10. Place where wine was bottled.
11. Trademarks.
12. Shipper's name.
13. Importer's name.
14. Liquor store name or distributor name.
15. Grape variety.
16. Diplomas, certificates, and classifications.
17. Ingredients added to the grape juice or wine.
18. Advertising messages, illustrations, mottos, and claims.
19. Product information.
20. Corporate name of the wine's producer.
21. Marks descriptive of quality.
22. Sugar content of grapes or wine.
23. Sulfite content.

Only the name of the wine (or kind), the percentage of alcohol, the bottler, and the place where the grapes were grown and crushed are mandatory on all wine labels.

Most of this information, when it appears, is on the main label of the bottle. However, some of it may appear on a seal affixed to the capsule of foil or plastic covering the cork, on a neck label, or on a back label.

Name of the Wine

The name on the wine label can identify the wine in the bottle as a unique wine, or it may mean very little. If the label shows "Château Mouton-Rothschild" as the name of the wine, almost the whole story is told: There is only one Mouton-Rothschild. The name identifies the vineyard, how it was produced, the microclimate, its consistency from year to year, and so on.

On the other hand, if the bottle says Liebfraumilch, which is a legitimate but invented name, little is learned. It is a German Rhine wine, but from which vineyard? How was it processed? Is it dry? Semidry? Much more is left unsaid. Between these two extremes are several different kinds of names that offer different degrees of information.

Geographic Names

Mouton-Rothschild, and other estates and vineyards in Germany, France, Italy, and the United States, give their names to the wines they produce. These names become more meaningful as they refer to smaller and smaller areas. Mouton-Rothschild, for instance, may be identified by a series of geographic names, each with fewer referents: French wine, Bordeaux, Médoc, Pauillac, Château Mouton-Rothschild. The geographic names have thus been narrowed from a country to a tiny plot of land. Likewise, in the United States, California is a geographic name, as is Napa Valley, with stags Leap or Carneros as more specific areas within Napa.

When large areas, countries, and whole growing regions are mentioned on the label as the place of origin, it says little about the taste of the wine—the territory is too large to know which microclimate the grapes were grown in, or anything about which winemaster made the wine or the production and aging methods used.

The middle ground offers more information but can be misleading. For example, Champagne, Burgundy, and Sauternes are places in France making very particular types of wine from specific grapes. Placing names such as "California Champagne," "Chilean Burgundy," and "Spanish Sauterne" on a label can be confusing—is it a wine that resembles the wine from these places, or would the producer like the customer to think so? For sure, the microclimates are not similar. Are the grapes used the same? Or only similar in taste? Or only alike in color?

French wines, when they are guaranteed as to origin by the government, come from a precise geographical location. Other countries are a bit more lax, permitting a producer to use a name taken from any place within a 15-kilometer (or similar) radius. Understandably, growers tend to use the most famous and successful name they can.

Varietal Names

These wines indicate that the wines contain a certain variety of grape. Assuming that the variety of grape, as named, is not misleading (a grey Riesling versus a true Riesling), buyers who are familiar with the grape (from experiencing another wine made from that grape) will have a good basis to judge the approximate taste of the wine. For example, someone who knows that Beaujolais contains Gamay grapes will recognize the California Gamay as having a similar taste, although the grapes are somewhat dissimilar due to the different climate and soil.

Generic Names

Some place names have become so widely used away from the actual location that they have a supposed identity all their own: Sherry, Madeira, Port, Burgundy. Unfortunately, little faith can be placed in a generic name, because its definition is too encompassing. True Sherry is made in a very time-consuming, precise process, resulting in a sweetish high-proof wine with a nutty cooked flavor. However, all sweetish high-proof wines with a nutty cooked flavor are not true Sherry. Likewise, Burgundy often refers to a dry, red wine made predominantly from the Pinot Noir grape in the region of Burgundy, France (although there are white Burgundies). But not all dry red wines are Burgundy, regardless of what the label says.

Brand Names

The producer may make up unique names that cannot be copied by anyone. They tell the entire story, or very much of it. Bottles of Bristol Cream or Dry Sack Sherry, produced by Harveys and Williams & Humbert respectively, are as surely those wines as a brand-name coffee is that brand.

Brand names of Champagne such as Bollinger, Krug, Perrier-Jouët, Veuve Cliquot, and Taittinger need very little further qualification—perhaps only degree of sweetness (e.g., Brut or Semi-Dry) and vintage, if any—to become an "exact" wine.

Country of Origin

The name of the country of origin qualifying a generic name can tell the purchaser a great deal. In France, Sauterne is grown in a specific region, from a certain grape, by a specific method of cultivation, under controlled yield per acre, and so on. Sauterne in California may be grown under as rigorous but different regulations, or it may be any slightly sweet white wine made in California.

Country of origin, if not the country the purchaser resides in, indicates that the wine is imported, having been transported across a border. If it is a very good wine and has traveled some distance (e.g., over an ocean), it will need to rest before being drunk. Wines get "motion sickness," being shook up during the journey and need to mellow on arrival. Considerable handling and import taxes may be included in the price. However, depending on the exchange rate of the currencies of the countries, imported wines may be a bargain.

Color

Sometimes the bottle will state the color of the wine as "red" or "white," which is useful information when the pale or dark relative of some famous wine is being considered. The white Beaujolais or the red Graves may be a bargain, but not if red Beaujolais or white Graves is used as the standard.

Year Grapes Were Crushed

The main wine label or a separate label may have a year printed on it. This indicates that most or all of the grapes used to make the wine were produced in that year.

"Vintage Year" can mean much more. Some growers do not declare a "vintage year" every year. Therefore, if it is a "vintage year," it means that, in the opinion of the producer and professional or government evaluators, that year was an excellent year. A vintage year is a product of a season that produced excellent grapes from which first-quality wine was made. When the year is poor, the particular producer may choose to sell the grapes or wine to a bulk producer, or sell the wine as a nonvintage year wine. The wine produced, even when made entirely of the grapes of one year, may be offered without an indication of the year. In some cases, the producer, usually of extremely prestigious wines, may not even offer the wine under the estate or vineyard name. It may be sold with only a geographic name.

It is important to understand the label laws of various countries. Whereas in France, vintage years connote excellence, Germany, for example, has a "vintage" every year; it is no indication of quality, only of the year the grapes were produced.

A bottle of wine with a production or vintage year on it that was not a great vintage year for the wine in question will probably still be priced somewhat higher than a comparable nonvintage wine. The nonvintage wine may actually be the better value if skilled winemasters have blended wines of several different seasons' production to make up for the deficiencies of the wine of the particular season.

Gauging a wine by vintage charts, which evaluate every year's wines by region, is unfair to many wines within the region; the charts necessarily represent colossal generalizations. Also, they tend to increase the price of bottles labeled with the year in high-rated years, thereby creating bargains of the nonvintage wines of that period and of the good wines in supposedly bad years.

Region Where Grapes Were Grown and Crushed

An indication of the region where the grapes were grown and crushed provides details that can be helpful, especially for generic or varietal names. California Sherry is not New York Sherry, nor is California Riesling the same as New York Riesling. The microclimate of the region where the grapes were grown will suggest whether they were "hot climate" grapes or "cool climate" grapes, and so on.

Kind of Wine

The label may indicate the kind of wine, such as table wine, dessert wine, aperitif wine, fortified wine, special natural wine (mixed with fruit juice), carbonated wine, sparkling wine.

Percentage of Alcohol

The alcohol content of the wine is given as a percentage on bottles of wine. On bottles of spirits, it appears as a "proof" equal to twice the percentage (12 percent equals 24 proof). Aperitif wines and Sherries have 17 to 20 percent alcohol; table wines, 7 to 14 percent (the fuller-bodied wines being 12 to 14 percent); dessert wines, 14 to 21 percent.

The percentage of alcohol can vary plus-or-minus 1½ percent from the stated percentage; use it as an approximation.

Liquid Ounces in Bottle

The exact contents of the bottle will be stated. A standard bottle holds 750 ml or 25.4 ounces, a liter bottle contains approximately 33.8 ounces. Some exotic or long-necked bottles may contain less than they seem to hold.

Official and Governmental Guarantees

Some wines have a potential market that far exceeds the production capacities of the region producing the wine. To guarantee that the wine in the bottle is as represented, the governments of several countries, notably France, Germany, and Italy, authenticate the origin of the wine.

To some extent, they also guarantee that the wine meets certain standards of excellence. For example, in France, where regulations are strict and well enforced, there are three *Appellations d'Origine,* which in addition to guaranteeing origin are tantamount to broad ratings. *Appellation Contrôlée* is the highest grade, to which almost all great French wines belong. The others are *Vins Délimites de Qualités Supérieure* (V.D.Q.S.) and *Appellation Simple* wines, which are seldom exported.

On a wine label, the name of the wine may be followed by the words *Appellation Contrôlée,* or by a phrase with these words

bracketing a geographic name; for example, *Appellation Pauillac Contrôlée*. The smaller the region named in the middle, the better the control, and often the wine.

The Italian wine law is similar. When the words *Denominazione di Origine Controllata* appear on a wine label, the wine is certified as to place of origin, grapes used, planting of vines, and so on. There is also a category above this one, *Denominazione di Origine Controllata e Garantia*, which is a quality standard given to selected wines that already merit *Denominazione di Origine Controllata*.

German wine law is relatively recent. The 1971 bottles, which arrived in the United States in 1972, were the first to use the new labeling system. There are three categories of "quality" German wines, of varying quality:

1. *Deutcher Tafelwein* ("German table wine"), which must be produced from approved grape varieties in one of five major table wine regions: Mosel, Rhein, Main, Neckar, or Oberheim. No vineyard name appears on these bottles.

2. *Deutscher Qualitätswein bestimmter Anbaugebiete (QbA)* ("German quality wine of designated regions"), which must be made of approved grape varieties, have at least 7 percent alcohol, and come exclusively from one of the 11 quality German wine regions. Every bottle carries a control number showing that it has been tasted and analyzed to ensure that it is worthy of the label. It can be labeled by the region, subregion, or vineyard in connection with the name of a village. To use its name on a label, a vineyard must be at least 2.5 acres in size; this eliminates 23,000 to 24,000 vineyards and simplifies the understanding of German wine labels.

3. *Deutscher Qualitätswein mit Prädikat (QmP)* ("German quality wine with special attributes") is the highest category of German wine. It is made from approved grape varieties and contains at least 9 to 10 percent alcohol without sugar added. Within the category, several subcategories—the attributes (Prädikats)—are indicated on wine labels:
   - *Kabinett* ("cabinet") wines must be made from fully mature grapes and have a fruity quality. They are the driest of the wines with special attributes.
   - *Spätlese* ("late harvest") wines are made from grapes picked after the completion of the normal harvest, when they are more mature. These wines have a richer, sweeter taste.
   - *Auslese* ("selected harvest") wines are made from the ripest bunches of grapes and are individually selected and pressed. Some of them have noble rot (*edelfaule*, or *Botrytis cinerea*),

which concentrates their juice and sweetness. They are sweeter and richer than Spätlese.

- *Beerenauslese* ("hand-picked selected harvest") wines are made from individually harvested grapes that have noble rot. These wines are sweeter and richer than Auslese wines.
- *Trockenbeerenauslese* ("shriveled hand-picked selected harvest") wines are made from noble rot grapes, harvested individually, which have been allowed to dry on the vine, shriveling and appearing like raisins, concentrating the sweet grape juice and flavor. They are very expensive and very sweet.
- *Eiswein* is made from Beerenauslese or Trockenbeerenauslese grapes that have frozen on the vine due to cold weather and have been quickly harvested (the night of the frost) and pressed (early that morning or at dawn). Once the grapes thaw, they become mushy and Eiswein ("wine from ice grapes") cannot be made.

Place Where Wine Was Bottled

Wine need not be bottled where it was made, any more than grapes need be pressed where they were grown. Some wines are sold to shippers in the casks where they have been aged and sometimes blended before being bottled under the shippers' control. Many less expensive wines are transported in tanker ships, like oil, and pumped from vineyard to ship to bottler.

A premium price is attached to the words "estate bottled," or the equivalent: *Mis en Bouteille au Chateau* or *Domaine* or *originalabfüllung.* This is a guarantee that the winemaking process has been controlled throughout by the property that produced it.

The question is really, "If a wine is made and bottled where the grapes were grown, is that alone a guarantee of any particular quality standard?" Vineyards of quality wines have established reputations; they guarantee production according to *their* standards; their names are known as symbols of quality. When buying from unknown or relatively new vineyards without an established reputation, estate bottling provides little assurance of quality. If the wine is good, the purchaser can begin to follow that vineyard; if not, it is best to forget that vineyard and try another. Also, if the wine is nonvintage, for example, the blend of several years, it is entirely possible that an experienced shipper/bottler might be a better blender.

The English have had a long tradition of importing wines in casks and bottling them in England. This has resulted in cheaper wines for the English customer, and often superior products have been saved from mediocrity by experienced cellarmen who

know how to blend mediocre wines into a well-balanced drinking wine.

Large American producers, such as *Gallo* and *Mondavi* have vineyards located throughout California. The wines from several of their vineyards are often blended to create their famous jug wines—good drinking wines at reasonable prices. The wines may be blends of wine from cooler regions, warmer regions, coastal regions, and central regions. The place where the wine was bottled may have very little to do with where the grapes were grown. However, the winemaster strives to blend wines from different areas and of different years, maintaining the same taste from year to year. Thus, a consumer who knows the taste of that wine from that winemaker has a standard of reference from which to judge future wines from the same winemaker.

Trademarks

Many wine names are trademarked to protect them against exploitation by other firms. An insignia, the artwork on the label, and indeed the entire label may be protected. Trademarks, however, do not reflect the quality of the wine, no matter how elegant they appear or how grandiose they sound.

Shipper's Name

Shippers often specialize in a quality of wine. Knowing what level of quality a shipper specializes in, and seeing the shipper's name on a product indicates that the wine meets its standards. Buyers can then purchase wines from the shippers whose products have been most enjoyable.

Importer's Name

Although in the United States importers do not usually blend wines themselves, most offer a somewhat consistent variety of wines. They usually define a market to target, then import wines for that market. Certain houses offer consistently good wines.

Liquor Store Name or Distributor Name

Many liquor store owners and distributors trading with restaurants are skilled specialists and offer wines that have been properly stored and are of good quality. These establishments also usually employ knowledgeable sales representatives, who can provide information on individual wines and can often supply wine vintage charts, brochures on various wines, and even recipes for consumers.

Grape Variety

The wine may prominently state the grape variety on the label, (e.g., Cabernet Sauvignon, Pinot Noir, Chardonnay) or it may not. Even when wine is not named after a grape variety, the label may indicate that the wine is made from a specific variety or varieties. Since the producer has no obligation to supply this information, the grape name on the label should be taken as a positive indication of quality.

Diplomas, Certificates, and Classifications

"Gold Medal Winner" and other such labels appear on some wines, indicating medals won in contests or at fairs. These can be noted, but taken lightly, for a great wine one season is largely irrelevant; each year, the climate is different, the handling of the grapes is different, and thus the wine is different. However, if a winemaker consistently receives medals, year after year, it indicates a serious, top-quality winemaker whose reputation is on the line—note those wines, purchase them, and enjoy them.

There have been efforts at certification of wine, for example, by the Association for the Development of the Exportation of Bordeaux wine (ADEB). This certification may appear on the wine label. (The ADEB used a seal on the capsule.)

A system of classification and certification does prevail for some of the great Bordeaux wines, labeled as a "first growth" "second growth," "third growth," "fourth growth," or "fifth growth" wines. The first growths (grand cru classe) are regarded by growers, winemakers, wine regulatory agencies, and connoisseurs as the best vineyards in that region and consistent over time. If they do not maintain standards of excellence, they can lose their prestigious rating. Some wines were classified in 1855, others in 1953, and still others in 1955. Some Bordeaux wines comparable to those classified (the Pomerols) have not been classified. Chapter Fifteen lists specific wines representing each growth. The chart in Chapter Eleven lists some of the best-regarded wines in the world; many of them are rated as "first," "second," and "third growth" Bordeaux wines.

Ingredients Added to the Grape Juice (Must) or Wine

When ingredients have been added to a wine, the addition must be noted on the label.

Sugar is sometimes added to the must produced in cold or rainy seasons. When the season has had too few warm, sunny days,

the grapes do not ripen; thus, they do not sweeten sufficiently, and there is not enough natural sugar for the yeast to feed on to convert the sugar to alcohol. Since an alcoholic percentage of about 12 to 14 percent is desired, additional "food" for the yeast (sugar) may, in some instances, be added to ensure a good wine, with a reasonable alcohol content.

When sugar is added to the must (grape juice), no mention of this is made on the label. Sometimes, the absence of additional sugar is noted on German wines by stating that the wine is natural (*naturwein, naturrein,* or *ungezuckert*). The most recent legislation restricts the practice of adding sugar, which is limited in any case to inexpensive German wines.

In California, due to its multitude of warm, sunny days, sugaring is illegal—also unnecessary if the winemaker is competent and experienced. In New York State, where cool summers and rainy falls are the norm, it is sometimes necessary to add both water and sugar to the must to increase the alcoholic content and to reduce the acidity. These additions are not noted on the label.

Advertising Messages, Illustrations, Mottos, and Claims

A producer or shipper can place any advertising message on a label. Although subject to fraud legislation of the areas in which the wine is produced or sold, extravagant praise for the product is allowed. For example, any wine may be called a great wine or grand vin. Other terms that may have some meaning are not rigidly enforceable; for example, première cuvée means the wine is among the best in an area, but it might easily be confused with Premier Cru, a classification that is rigidly controlled.

There is a good deal of "quality by association" promulgated by wine labels. For example, château bottling is a selling point, indicated on wine bottles by some phrase on the order of *Mis en Bouteille au Domaine*. Other similar phrases appear on labels when in fact the wine was bottled away from the place where the grapes were grown. For example, *Mis en Bouteille dans nos Caves, Mis en Bouteille dans nos Chais,* and *Mis en Bouteille à Fuisse* mean that the wine was not bottled at the estate where the grapes were grown and the wine produced. In one instance, a wine was bottled at another château and this fact was proudly proclaimed.

The reputation of the great Bordeaux estates, enforced by a system of classification, has encouraged producers in other areas to dignify their labels with a picture of a castle. Even in Burgundy, wines that have the name of a particular château exploit this association, when in fact there is no classification of Burgundy châteaux and other words in smaller type on the label are more important, specifically the vineyard area.

Sometimes, actual names are exploited by near spellings; for example, Château Lafite has occasioned some Laffites and Lafittes.

Product Information

Product information includes advice on how to serve the wine—for example, "well chilled," or recommendations of food dishes that are complemented by the wine—or even how to cook with it. The producer may also choose to discuss such matters as the location of the vineyard, the history of the area and the wine, famous people who have enjoyed the wine, or the heritage and history of the winemaker's family.

Corporate Name of the Wine's Producer

The label states the company name that owns the property on which the wine was grown and produced. This is easy to locate on American wine labels but somewhat less clear when written in French, Italian, or German. On French labels, for example, the abbreviation *Ets.* (for establishment) indicates the company. *S.A.* on French, Italian, and Spanish labels is the equivalent of "incorporated."

Marks Descriptive of Quality

Sometimes wines, such as Cognac brandies, are labeled with initials that may have some meaning; for example, Marsala wine may have the letters C.O.M. (Choice Old Marsala) or S.O.M. (Superior Old Marsala). On French wines, a virtually meaningless C.E. (cuvée extra) may appear.

Sugar Content of Grapes or Wine

In addition to the German wine terms discussed earlier, Champagne provides another notable example of a wine's sweetness being indicated on the bottle.

When Champagne is being recorked after the second fermentation has taken place in the bottle (to give it the bubbles), a small amount of sugar syrup and wine is added to replace the sediment from the second fermentation, which has been removed.

The quantity of sugar syrup added to the Champagne determines the sweetness of the wine. This is indicated on the bottle. *Brut* indicates up to 1½ percent added sugar-wine, *Extra Dry* up to

3 percent, *Dry* or *Sec* to 4 or 5 percent or more. The words *Gout Anglais, Gout Américain,* and *Gout Français,* after the major markets for Champagnes of varying sweetness, correspond to Brut, Extra Dry, and Sec, respectively.

Sulfite Content

If the wine contains sulfites comprising more than 10 parts of sulfites (especially sulfur dioxide) per million parts of wine, the label will state, "Contains sulfites."

Price and Quality

In buying wine, purchasers usually get what they pay for. A renowned wine that sells at a substantial price should be a great wine. A person who buys a mediocre wine in an outlandish bottle gets a mediocre wine in a great bottle. The purchaser buys not only the liquid in the bottle plus that much glass and that much cork but also a reputation, real or fancied, deserved or unmerited. The greater the reputation, the greater scarcity of the wine, the greater the demand for it, the greater the price. That is why one wine costs a hundred or two hundred times more than another. The cost of manufacturing, bottling, merchandising, and shipping a great bottle, with the exception of the late harvested wines, is about the same as the cost for an inexpensive wine.

The bargains in wines are those that do not have the reputation of the higher-priced well-known wines, but have quality and character to commend them. Many American wines are bargains because they are from relatively young vineyards trying to get established. Bargains in the French wines include those bearing an Appellation d'Origine; they are surely among the V.D.Q.S. wines that are not yet "in." The bargains in the Bordeaux wines are those that have not been classed or are classed below the well-known crus, such as the wine classed *Crus Bourgeois Supérieurs* and *Crus Bourgeois.*

Certain wines are definite nonbargains. Inexpensive French, German, Italian, Spanish, and Portuguese wines have to be transported great distances, handled by innumerable middlemen, and taxed by the federal government before they can be sold. How much of the few dollars they bring can be payment for the wine? On the other hand, a California wine in a jug has a much lower overhead per ounce and may be a better value.

The greatest values, however, are achieved by buying whatever your market enjoys, at whatever price you feel the wine is worth—an individual price/performance test for the wine is the best approach. An expensive dry, full-bodied Bordeaux or Cabernet Sauvignon to a person who enjoys semidry wines is not a good value. Know your customers, what they enjoy, and purchase for them.

Tasting, Storing, and Drinking Wine

Knowing how to read a label is important; having "book knowledge" of wine and winemaking processes is also important, but you must taste wines to really know and understand them. How to taste and evaluate wines is a learned skill. There are systematic steps to tasting wines that are generally followed at any organized wine tastings or tastings/seminars. These tastings are an excellent way to become familiar with wines.

In addition to learning how to taste and evaluate wines, it is essential to learn how to care for wines you have purchased. How should they be stored? At what temperature? At what level of humidity? Near the sunlight or away from it?

Once you know how to taste wines and have purchased and stored several cases of various wines, you now need to know how to use them appropriately for different kinds of occasions and with various kinds of foods. This chapter provides guidelines for tasting, storing, and using wine.

❖ GUIDELINES FOR TASTING AND EVALUATING WINES

Most wines are "sound"; they are drinkable, not overly acidic or vinegary, and not poorly made. One sound wine is not better than another. The opinion of the person drinking the wine is the only real standard to use: if the individual likes it, it's a good wine. If the person doesn't like it, it's so-so or a bad-tasting wine, but not a "bad" wine unless it has spoiled.

There are, however, some basics that everyone can use to judge wines—some things to look for, to taste for, to evaluate.

Wines have remarkably different qualities; evaluation requires (1) knowing what to look for in wines and (2) knowing what to expect of a particular wine.

By knowing what to look for, you can establish benchmarks for evaluation. For example, the color of the wine is an important aspect of its quality. Every evaluation should include a consideration of the wine's color. Although there are no "right" colors for wine, there are proper ranges for a particular grape, and definite "wrong" colors for certain wines.

When evaluating a wine, as opposed to just enjoying it, a stricter set of guidelines is necessary.

Selecting Wines for Evaluation

An easy way to evaluate wines is to attend a tasting arranged by a purveyor, importer, or wine society—or you might even purchase the wines to evaluate them.

At a particular session, the wines tasted should be related so that each wine's particular qualities are sharply focused; for instance, perhaps six sparkling wines, six dry red wines, six dry white wines, or six dessert wines. It is best to avoid dramatic contrasts; the tastes can be so different that the dry wines will taste bitter or acidic compared with sweeter wines. The sweetness of a wine can alter the palate and distort the true taste of dry wines.

During a wine tasting, progress from the less complex wines to the more complex and from the dry to the sweet. In other words, minimize the influence of the wine tasted first, using that as a sort of "base" or "standard." The tasting of young wines before mature wines, and dry wines before sweet wines, regardless of color, helps to protect objectivity.

Evaluate no more than six or seven wines in a single session. If you are serious about the wine tasting, you should not swallow the wine but spit into a cup or spittoon so the alcohol is not absorbed into the circulatory system. After tasting and swallowing a few wines, the alcohol can impair the taster's judgment.

Setting Up the Room for the Tasting

It is very important that the setting not intrude on the tasting and evaluation of wines. Ideally, the evaluation should take place in a modest room, lighted by daylight, with a temperature of 68°F and a humidity of 50 percent. Obviously, a marvelously contrived cellar scene can lend a very ordinary wine credentials it does not deserve. Daylight shows wine as it really is; normal incandescent light is acceptable, but colored lights and fluorescent lights distort a wine's appearance. The temperature and humidity of the room affect the taster's abilities. A very light, well-chilled wine may appear extraordinary in a hot room, while an excellent heavy red may only add to the taster's discomfort.

A white cloth or sheet of paper should be provided for proper color evaluation; windows with open sky behind them or a white backdrop, such as a white wall, are also acceptable. A candle or weak light bulb permits an examination of clarity.

Personal physical conditions may potentially influence tasting abilities. For example, head colds and allergy attacks severely distort taste perception, sometimes sensitizing particular qualities, sometimes masking them.

Smoking while tasting wines distorts taste perception and should be refrained from during the tasting and immediately beforehand.

Foods eaten during or just prior to the tasting can affect the tasting. Cheese makes wines taste exceptionally good; fruits, due to their acidity, may make certain wines taste bad.

Condition of the Wine

Wines being tasted should be offered in a condition similar to that in which they will normally be served. Briefly, red wines should be at room temperature and white wines and sparkling wines should be chilled. Red wines should be opened prior to tasting, and any wine should have the opportunity to rest from shipping or handling. Old wines with sediment should be decanted.

Certain aspects of the wine should be considered. For example, some inexpensive white wines have an excellent bouquet (aroma) when they are first opened, but lose it after a few minutes. A restaurateur buying wine for a restaurant in which people linger over a bottle would consider the durability of the wine's attractiveness an important purchase point.

The taste of wine changes with temperature variation. An acceptable white wine at 40°F may be undrinkable at 70°F. A young red wine that is unacceptable at room temperature may be quite pleasant when slightly chilled. If the purchaser anticipates serving the wine at a temperature other than normal serving temperatures, that variation should be duplicated at the tasting.

Glasses

The standard wineglass, a stemmed 8-ounce tulip glass, should be used rather than any specialty or decorated glass. The glass must be absolutely clean and well polished. Ideally, it will be of crystal (glass that includes lead compounds in its formation) and not have the reinforced, slightly curled edge called a safety rim. A separate glass should be provided for each wine.

Special "tasting" glasses are available and may be used if the participants so desire. These glasses are scientifically designed to bring out the true qualities of the wine. A popular one is flat-bottomed with an indentation in the center of the bottom, and sides slopping inward to capture the bouquet and aroma (see Figure 11.2).

If purchasing wines for a restaurant, and a tasting was done with a special tasting glass, the procedure should be repeated with the glasses used in the restaurant because the tasting glass will make the wine taste better (or worse) than when tasted in customary glassware.

The Tasting Procedure

Disciplined, orderly tasting is necessary if the wines are to be judged fairly and accurately. Note taking is a necessity when tasting related wines.

There is a standard procedure for tasting and evaluating wines: examine appearance, bouquet, and taste, in that order.

What to Look for When Evaluating Wine

Appearance

There are five aspects for judging the appearance of wine:

1. Color.
2. Depth of color or intensity of color.
3. Clarity or brilliance.
4. Deposit or sediment.
5. Viscosity or fluidity.

Color

Although wines are often described as red, rosé (pink), or white, they vary tremendously within these categories. White wines may range from a nearly colorless watery white to a dark amber, with

PROCEDURE FOR TASTING WINE

Appearance

1. Pour an ounce of wine into the wineglass.
2. Pick up the glass by its stem (if there is one) and hold it at eye level, preferably between the taster and a window. Examine the wine, looking for clarity and brilliance.
3. Hold the glass, still at eye level, between the taster and a candle or weak light source. Slowly rotate the glass. Tilt the glass to one side and look at the meniscus, or surface of the wine against the air, noticing any tinge of green (youth in white wines), amber (age in white wines), purple (youth in red wines), or brown (age in red wines).
4. Holding the glass by the base, swish the wine so that it washes up and down the walls of the glass. Observe drops draining back into the glass—the legs. Note the length of time for a drop to travel down the side of the glass.

Bouquet

5. Holding the glass by the base, swish it again to expose a large amount of wine to the air to release the bouquet.
6. Hold the glass directly under the nose; ideally, the upper half of the glass should be over the taster's nose. Inhale the bouquet.
7. Warm the wine slightly by cupping the base of the globe in the hands, and repeat step 6.

Taste

8. Take a small amount of wine, perhaps a teaspoonful, into the mouth. Allow it to pass over the tongue and palate (roof, sides, and back of mouth).
9. Purse lips and inhale some air to gather the various elements of the wine bouquet and to drive them into the nasal passages, permitting an appreciation of the wine aroma.
10. Spit out the wine.

overtones and reflections of green, yellow, straw-yellow, and brown apparent in many white wines. Rosé wines vary from a true rose, a very light pink, to a rich cherry color. Wines that are described as red may in fact be ruby, garnet, purple, violet, brick red, dark cherry, or orange.

Color is a valuable point of reference in two ways: (1) Individual wines have characteristic colors that can be noted for reference; when they are "sick" or adulterated, the color will vary; and (2) some objective observations are possible from an examination of a wine's color once the taster has had some experience in evaluation.

In red wines, a genuinely purple color, especially in a wine that is not ordinarily purplish, indicates immaturity. Wine in the cask is purple. As the wine matures, it loses its purple color and becomes closer to a ruby red. If a bluish color is apparent in a supposedly mature wine, it probably means that the wine has been adulterated.

Older red wines may have a red-brown or even amber-brown coloring, especially apparent at the point where the wine touches the glasses when the glass is tilted. While this coloring is characteristic of mature wines, it should not be present in young wines. Premature browning usually indicates that the wine has been overheated during fermenting or storage and has become oxidized.

Unlike red wines, which gradually lighten in color with age, white wines become darker. For example, a sweet wine is more yellow when young. After maturing it becomes golden.

A touch of brown in a young wine indicates that it will not keep well. A definite brownish tinge in an older wine indicates that it is mature, most likely overmature or oxidized. When a white wine has become oxidized, it takes on the color of Madeira— a definite yellow-brown—and is described as "amaderized."

Rosés of quality have a definite pink color that departs significantly from the light red of watered-down red wine. A little orange is acceptable, especially in rosé from hot climates, but excess orange color is not. Purple and blue notes in a rosé signal deterioration or improper manufacture.

Depth of Color

Depth of color requires some experience to analyze. There is a definite difference between a ruby-red wine and a deep ruby-red wine, although in each instance the color is only describable as "ruby red." A high color intensity indicates that the grapes used are of excellent quality, ripe, fully developed, and well nourished, and that the wine was properly processed. A pallid color, although the same color, indicates a poor year or an accelerated processing.

Clarity

When a wine is held up to a candle or a weak lightbulb, or placed in a silver saucer-shaped cup with a pebbly bottom, called a "tastevin," the glints of light passing through it show its degree of clarity and highlight any floating or suspended particles.

The wine should be clear, bright, and somewhat shiny. Haze, cloudiness, and dullness are definitely bad signs. The evaluator should not be misled by bits of cork or some floating sediment. Neither reflects on the quality of the wine, merely its handling. Likewise, wines that have recently been shipped or subjected to radical changes of temperature may be momentarily clouded and this should be considered.

Deposit

Deposit or sediment is more apparent in the bottle than in the glass. As wines age, they may throw off certain compounds, specifically pigments, tannins, and mineral salts, which settle to the bottom of the bottle, either immediately if they are heavy, or after some time if they are "flaky." While sediment is not necessarily a sign of a great wine, it is characteristic of many mature excellent wines and should not be an immediate cause for rejection.

Viscosity or Fluidity

When the wine is swished in the glass almost to the rim, it may either fall back into the base quickly, or cling to the sides of the glass and fall back in "tears" or "legs." This heavy transparent film is indicative of richness and is quite prized. It also indicates alcohol content (the slower the drop returns to the wine, the higher alcohol content) and sweetness (the sweeter, the slower the drop travels).

In addition to these factors, sparkling wines might also be judged for the fineness of their bubbles and the longevity of the evanescence.

Bouquet

Each wine has a characteristic bouquet or "nose" that comprises two elements: (1) the aroma characteristic of the grapes from which the wine was made; and (2) the perfume, which is the smell acquired during fermentation in the vat, aging in the cask, and maturation in the bottle. In a young wine, the aroma is strong; gradually the perfume dominates.

Objective assessment by smell is not directed to the evaluation of a specific bouquet, which the evaluator may or may not like, but to intensity of the bouquet and its cleanliness.

The bouquet of wine can vary in intensity from full or most intense to evasive or least intense. In general, the following terms describe descending intensities: full, pronounced, delicate, subtle, light, and evasive.

Cleanliness or lack of cleanliness should be immediately apparent. The first impression counts, as the nose quickly becomes fatigued. Any "off smell" should be a cause for rejection, including smells resembling vinegar, rotten eggs, metals, wet rope, cooked cabbage, mold, ink, mustiness, or woodiness.

It is also possible that there may be an unpleasant "cork smell" because the wine's cork has decayed. Although the smell is unforgettable, "corked" or "corky" wines are relatively rare.

Taste

The more objective aspects of taste can be separated from less objective characteristics of wine or its flavor. The taste of wine is essentially limited to its sweetness, acidity, bitterness, harshness and roughness, richness or body, alcohol content, and the persistence of its aftertaste. Particular flavors, for example, a raspberry flavor, like a particular bouquet, characterize individual wines.

There must be sugar in a wine for it to taste sweet. If fermentation (the conversion of natural sugar in the grape juice to alcohol) has been complete, as is the case in most red wines, the wine will be dry, not sweet at all. White wines, which are more likely to contain some sugar, vary from extremely dry (no residual sugar) to very sweet (dessert wines with residual sugar). A high alcohol content or high acidity tends to mask sweetness, and vice versa.

The absence of sugar does not make a wine "acidic"; it makes it "dry." Likewise, a wine may be both sweet and acidic. Acidity, in other words, is a separate dimension of taste, not a degree of sweetness. Sweetness is the most apparent taste characteristic; acidity is the next most noticeable. Medium acidity gives wine character, while too much acidity can make wine taste sharp, tart, green, or make it undrinkable. Too little acidity can make it appear dull, listless, and flat.

Bitterness is another dimension of taste, distinct from acidity. Quinine is a common bitter substance, characterizing bar bitters, tonic water, and aperitif drinks. Some wine may have this taste (although it is not due to quinine), and some people may find it pleasant in moderation.

Harshness or hardness is due to the presence of tannin compounds and therefore is more characteristic of the taste of young red wines. Tannin is precipitated or mellowed as a wine ages. In fact, the primary reason for allowing a wine to age or mature in the

bottle is to give the harsh tannin time to mellow and blend with the fruitiness of the wine, fully exposing the qualities of the wine. Waiting permits the fruitiness, acidity, and tannins to become balanced in the bouquet and taste of the wine.

Three factors contribute to the body of the wine, or how "heavy" it feels in the mouth: the dissolved material in it; the alcohol content; and the glycerine that gives the wine its unctuousness. On these bases, a wine may appear to be full, flat, and heavy or slight, slender, and light. Depending on the individual wine and a person's taste preference, body may be a sought-after asset in wines.

Aftertaste or finish might be better described as the persistence of characteristic taste and flavor. Quality wines continue to offer the same taste profile as in the original tasting, and that taste remains on the tongue even after the wine has been swallowed. Poorly made wines change taste, literally in the taster's mouth, and disappear the instant they are swallowed.

Knowing What to Expect in a Particular Wine

Establishing Quality Standards

A quality standard is easily established for certain aspects of a wine's appearance, bouquet, and taste. Nobody likes a muddy wine with a musty smell and an "off" taste: That would be a bad wine.

A wine may be fairly evaluated by anybody in such terms as its intensity of flavor, pleasantness of bouquet, astringency of taste. The problem in evaluating wine is not in judging these aspects and rejecting the wines with dull appearance, little bouquet, and an unpleasant taste. The difficulty is in evaluating various wines that are acceptable in those respects. Why is attractive looking, pleasant smelling, pleasant tasting wine X worth ten times as much as attractive looking, pleasant smelling, pleasant tasting wine Y? Wine X has hard-to-define characteristics that make it the best of its kind, the rare wine that is distinguished by very subtle qualities.

A person who is aware of the essential aspects of ordinary and good wines of a kind will be able to recognize a distinguished wine. The taster should also be able to discern those qualities that characterize particular types of wine. For example, distinguished German Riesling white wine will have a fruity bouquet; a distinguished mature red Bordeaux may exhibit a deep berry "fruitiness" that is quite distinct from a Riesling.

The purchaser may also encounter "special" bouquets and flavors in particular wines that can be misleading if not expected. For example, the Italian red wine Bonarda d'Asti has a sweet aroma of violets that is likely to surprise the taster pleasantly, while the excellent French Bordeaux Petrus can have a slightly woody taste that may disturb a taster who does not expect it.

When seriously evaluating wines, the taster should have had a prior experience with that variety of wine at a tasting or in a restaurant and possess an accurate taste and smell memory. In addition, detailed notes and good note-taking ability are a "plus." It is helpful to research the particular characteristics of the wines to be evaluated and to discuss them with others prior to the actual tasting.

Wine Description and Terminology

The specific and particular characteristics of a wine's bouquet and flavor cannot be accurately described, except in those rare instances when they can be likened to another smell or flavor; for example, that of roses or strawberries. After an accurate description of a wine's objective qualities, even experienced tasters' notes deteriorate. "Typical" is the term most commonly used to identify a hard-to-describe bouquet and flavor. Some descriptive words have a more or less standard referent and are somewhat more helpful in communicating a wine's qualities without drinking the wine.

In addition to previously discussed terms, self-evident terms (e.g., "sulfury"), and terms that defy definition although they are widely used (e.g., "noble," "breed," "character," "finesse"), the following terms frequently occur in wine discussions and literature:

COMMON WINE TASTING TERMINOLOGY

Austere	The bouquet of young wines that have quality but at the moment are undeveloped and immature.
Balanced	Indicates all the wine's features complement one another well and form a harmonious whole.
Coarse	Said of a rough, raw-tasting wine that shows no promise of ever becoming smooth; a sign not of immaturity but of lack of quality.
Common	Drinkable and pleasant but nothing special.
Deep	Indicates the bouquet and flavor of the wine have numerous overtones; the flavor or bouquet is complex.

Dumb	Said of a wine that is promising but as yet undeveloped; the connotation here is mute, not stupid.
Flat	Lacking interest, insipid, probably because of insufficient acidity.
Flinty	A taste that refers to the smell of gun flint; a pleasant bite, a clean sharpness.
Foxy	A taste associated with grapes of the native American vines, perhaps best described (although unfairly) as slightly fetid.
Green	Said of wines that are not ripe, in the sense of green fruit.
Heady	High in alcoholic content.
Light	Indicates either low alcoholic content or a lack of body.
Meaty	Along with similar words (e.g., fleshy), means that the wine is rich in suspended and dissolved material and has a very discernible mouth feel.
Mellow	Said of wines that are smooth, mature, and ripe and have no astringency.
Penetrating	Indicates that the wine's bouquet quickly fills the nasal passages.
Piquant	A pleasant acid taste, not sharp but attractively tart, except of course when the wine should not be acid at all.
Ordinary	Similar to common, but less of a pejorative because no disappointment is implied.
Robust	Indicates the wine is full-bodied and very strong in its flavor statement but still has quality and no really bad features.
Rounded	The same as balanced or well-balanced.
Sound	Well-made, free of defects, but not something said of a great wine, which presumably has more to commend it than the absence of faults.
Stalky	Means, with other terms such as twiggy and stemmy, that there is an excess of flavor from skins, stalks, seeds, and stems.
Sturdy	Said of a wine that will age well.

❖ GUIDELINES FOR THE STORAGE AND THE CARE OF WINE

Why store wine at all beyond immediate needs? There are two strong arguments against large-scale, long-term storage: (1) Most wines are at their best immediately or relatively soon after bottling; improvement after a year in the bottle is the exception, not the rule; and (2) wine merchants are generally much better equipped to store and care for wines than individuals or individual establishments.

Many merchants and individuals have invested in wines that deteriorated during storage (they were at their prime when purchased) and that, by the time they were finally opened, could not be consumed or sold. Quality wines should not be warehoused; they should be purchased on a continuing basis, for immediate consumption and/or sale (if a merchant), in quantities that reflect needs.

The prudent course for both the consumer and the restaurateur is to transfer the storage problem and the risks to a merchant or purveyor. Storage by individual consumers and restaurateurs should be limited and short-lived.

There are three justifications for storage of bottles of wine: (1) a definite savings as opposed to a speculative investment, (2) convenience, and (3) "wine collecting," either as a personal hobby or as part of commercial image building.

In certain parts of the country, the restrictive liquor laws make it advantageous, indeed necessary, for both individuals and restaurants to purchase wine in quantity. Otherwise, they would be restricted to the few bottles offered by the state-controlled stores. In these instances, storing wine, provided the proper wines are chosen, can effect a savings, a wider choice of wines, and enjoyment of wines otherwise unavailable.

Convenience is another reason. The cellar should be sufficient to supply memorable-occasion wines as well as ordinary table-use wines, and the restaurateur should have in inventory sufficient selections and quantities of each to assure customers their choice. On the other hand, only a certain amount of wine can be justified on the basis of convenience as it must be consumed before it overmatures, and the cellar must be temperature and humidity-controlled (as well as access-controlled in a restaurant).

Wine collecting can be a worthwhile hobby, but the collector may be paying for this pastime in bottles that cannot possibly be drunk before they deteriorate. By the same token, the restaurateur who has chosen to establish a romantic and well-stocked wine cellar for the enjoyment of guests may have to assign some of the cost of lost wine to publicity, merchandising, or interior decoration. Often, in these circumstances, the worst thing possible happens: The connoisseur drinks wines on the edge of undrinkability as the restaurateur with the magnificent cellar pushes bottles that have begun to decline to minimize losses.

Personal Cellar Lists and Restaurant Wine Menus

The individual consumer, purchasing for personal consumption, and the restaurateur, purchasing for restaurant sales, have two almost identical concerns. They both want wines that can be consumed on a daily basis as a routine mealtime beverage, and they both want wines that will satisfy an occasional fancy for something rare, exceptional and exciting. To pursue the parallel, they both need a cellar that emphasizes ready-to-drink, pleasant, moderately priced wines but offers some exotic, more exceptional bottles.

It may be that both the consumer and the restaurateur will come to a similar conclusion in regard to the everyday wine: the purchase of quality domestic wines in gallons for use in carafes. The wines are good, pleasant, consistent, readily available, and inexpensive.

The cellars of better bottles can also be similar—only the total quantity of bottles will differ, not the representation of wines. In fact, the only difference between a modest personal cellar and an extensive one is the number of bottles that would be recommended and perhaps the emphasis in the modest cellar on types of wine rather than on wines of particular estates.

The chart on pages 272–274 shows a sample wine cellar collection based on 100 bottles. It also lists percentages of each type of wine, the total collection being 100 percent. A proportional representation of wine types and imported and domestic wines is apparent. An individual may want to scale down a personal cellar to less than 100 bottles, while a restaurateur may well convert bottles into cases depending on the volume of business.

The Wine Cellar

Traditionally, cellars were used for wine storage because they provided desirable temperature and humidity. With modern air conditioning, the wine "cellar" may be located anywhere and may

REPRESENTATIVE WINE CELLAR

Wine Classification (%)		Number of Bottles (%)
Basic Type		*Total*
Red wines	49	
White wines	45	
Rosé wines	6	
		100
Still wines	94	
Sparkling wines	6	
		100
Origin		*Total*
Imported		
French	48	
Non-French	17	
		65
Domestic		
California	30	
New York	5	
		35
		100
Varieties		*Total*
RED WINES		
French		
Bordeaux Appellation Contrôllée	6	
Classed Châteaux	8	
		14
Burgundy		
Beaujolais	2	
Gevrey-Chambertin	2	
Nuits Saint-Georges	1	
Volnay	1	
Pommard	1	
Vosne-Romanee	2	
Châteauneuf-du-Pape	1	
Côte Rotie	1	
Hermitage	1	
Échezeaux	2	
		14

Wine Classification (%)		Number of Bottles (%)
Basic Type		*Total*
Italian		
Barolo	2	
Chianti	2	
Barbaresco	1	
Grignolino	1	
Barbera	1	
Valpolicella	1	
		8
Other Imported		
Vintage Port (Portugal)	1	
Sherry (Spain)	1	
		2
Domestic		
California varietals	10	
New York varietal	1	
		11
Rosé		
California	2	
New York	1	
Tavel (France)	1	
Anjou (France)	1	
Provence (France)	1	
		6
WHITE WINES		
French		
Bordeaux Appellation Contrôllée	2	
Classed Châteaux	2	
		4
Burgundy		
Chablis	2	
Pouilly-Fuissé	1	
Montrachet	1	
Meursault	1	
		5
Other		
Alsace	2	
Muscadet	1	
Vouvray	1	
Champagne	2	
V.D.Q.S.	2	
		8

Wine Classification (%)		Number of Bottles (%)
Basic Type		*Total*
Italian		
Capri	1	
Soave	1	
Montefiascone	1	
Frascati	1	
Vérdicchio	1	
Chianti	1	
Orvieto	1	
Asti Spumanti	1	
		8
German		
Rhine	2	
Rhinehessen	1	
Moselle	3	
Nahe	1	
Franconia	1	
Palantinate	1	
		9
Domestic		
California varietal	7	
California Champagne	2	
New York varietal	2	
		11
		100

be any size from the specially manufactured closetlike cellars (temperature- and humidity-controlled) to vast commercial warehouses.

Physical Requirements for the Storage of Wine

Any area for the storage of wine should be:

1. Between 48°F and 64°F at all times.
2. Vibration free.
3. Sheltered from the direct rays of the sun.
4. Free of substances that emit strong odors.
5. Spotlessly clean.
6. Relatively dry, but not arid (relative humidity of 55%).
7. Ventilated.

The wines should be stored on racks or in bins off the floor of the cellar. Commercial racks are available, starting with modules that accommodate three bottles. Or bins may be constructed so that the wines may be stored at a slight slant to keep the bottom of the corks moistened. Ideally, the center of the air bubble should be in the center of the bottle.

Handling Wine Bottles

Highly processed wines, such as those sold in gallon containers, can be shipped and moved without any ill effect. Other wines are more vulnerable and can develop a temporary cloudiness after traveling. Wines with sediment normally have the sediment shaken throughout the bottle during transit. These wines should be rested before they are used.

Wines with sediment should always be handled gently. Ideally, the wine will remain undisturbed in an out-of-the-way place in the wine cellar until it is used. Store these bottles with the label facing up, so that by gently pulling the bottle out of the bin it will be possible to see which wine it is without turning the bottle and disturbing the sediment. Allow bottles to remain undisturbed; don't dust them, just leave them. When ready to use, gently pull the bottle out of the bin without twisting or turning and carry it to the decanting area just as it was stored, label facing up.

Decanting Wines

Wines that have developed a sediment should be decanted, or poured from the original bottle into a serving vessel, before they are served. Although the wine may be poured directly from the bottle into wine glasses, there is more waste that way, and it is likely that some sediment will get into the glasses. In addition to separating the wine from its sediment, decanting also accelerates the oxidation of the wine, bringing out its bouquet. Sometimes wines are decanted for this reason alone. Most of the time, only mature quality red wines must be decanted; but some varietals develop a sediment, and it is possible to find some in a heavy white wine. (See Chapter Eleven for detailed instructions for decanting wines.)

Record Keeping

There is little use in cellaring quantities of the same wines unless they are carefully followed as they mature. The individual consumer may use an ornate cellarbook, while the restaurateur may use less romantic but equally effective bin cards. The basic information each requires is the same:

1. Date of purchase.
2. Purchase price.
3. Seller (merchant's name).
4. Date of and initial evaluation, if the wine was tasted; if not, the catalog description, or a clipping of the article that prompted its purchase.
5. Dated Tasting Notes. As a bottle is used or tasted, a dated tasting note should be prepared.
6. Updates. Updated information on the wine, especially price changes, with dates of updates and source of information (e.g., Christie's Wine Auction) should be posted.
7. Recommendation for future purchases based on tastings to date and price/performance.

❖   # SERVING WINES

The object of serving wine properly is to enhance the wine drinker's pleasure. Details of correct wine service were presented in Chapters Seven and Eleven. This section of Chapter Fourteen will emphasize niceties that will bring out the wine's best qualities. Having purchased the wine, and perhaps having nurtured it, it would be unfortunate to miss the opportunity to enjoy it fully.

Preparing Wine for Service

Allowing Wines to "Breathe"

Wine in the bottle is biologically dead but chemically alive. It reacts strongly to the presence of oxygen, which is why so much importance is attached to the soundness of the cork during maturation. When a wine is to be drunk, additional oxygen chemically feeds the flavor and the bouquet. The wine loses its "storage" taste and smell.

There are no absolute rules for determining how long a wine should be allowed to stand opened before being served. It is a function of the wine itself: how complex and full-bodied it is, how it was stored, whether it was shipped very far or is still at the original château, and so on. Among the most important aspects of the sommelier's stewardship is a bottle-by-bottle decision on this matter. Experience with the particular wine is undoubtedly the best guideline.

Some general observations are possible. Wines with discernible carbonation are served immediately with no "breathing" time. White wines and very young wines generally profit by a half hour of breathing time. Rosés, with rare exceptions, do not. A mature red wine up to 8 years old realizes its full potential after 1 to $2^1/_2$ hours of breathing time. On the other hand, very mature wines (25 years old, for example) may be too delicate to be "traumatized" by an onrush of oxygen. Often they must be served immediately, or their greatness, their delicate bouquet and flavor, wanes. Between 8 and 25 years, a proportionate breathing space is indicated. It is better to underexpose than to overexpose the wine to air, as the wine can breathe further in the wineglass.

Also, storage of wines in a nonoptimal setting (where the wine cellar is above 55°F) will quicken the pace of the wine's maturation in the bottle, so less breathing time will be necessary.

If you are unsure how much breathing time, if any, a bottle requires to reach its full potential, open a bottle and taste a sip every hour, recording tasting notes. When the wine has optimum flavor and bouquet, note the breathing time. Use this breathing time for other wines of the same case(s) purchased at the same time and stored in the same manner as the test bottle for the next 6 months or so. Then test a bottle again. The needed breathing time will gradually decrease as the wine ages.

Allowing a wine to aerate, or breathe, is also an excellent technique to age a young wine, particularly one with harsh tannins and that is a bit acidic. Open it early in the day and allow it to breathe for 6 to 8 hours to aerate and take the edge off the wine. Taste the wine every hour and if it is ready to drink before time to serve it, recork it using a special "closure cork" and pump out the air using a special pump designed for the purpose (e.g., "Vin-u-Vac" or other brand).

Serving Temperature of Wine

Ultimately, the temperature at which a wine is served is a matter of taste. Different people appreciate different aspects of a wine's quality as they are revealed at different temperatures. However, the majority of people—amateurs and professionals alike—agree on some general guidelines. The way the wine reacts to various temperatures determines the basic rules.

The lowest acceptable temperature for drinking wine appears to be about 36°F for very dry sparkling wines; the top limit for red wines is about 72°F. Below 36°F, the wine risks having most of its character frozen. Above 72°F, undesirable flavor elements dominate.

Most white wines are chilled to about 45°F, but sparkling, high-alcohol, or very sweet wines can be cooled to 42°F. It should be remembered that once the wine is poured, its temperature climbs rapidly and overchilling is quickly rectified.

Most red wines are best savored at 65°F. Red wines without much body or with unpleasant roughness profit by chilling.

Ideally, wine is chilled (or warmed) slowly in a room or area of the proper temperature. It can also be chilled satisfactorily in a refrigerator. A freezer compartment may cause it to "break," or precipitate tartaric acid crystals, by chilling it too quickly. Ten minutes in a bucket filled with crushed ice and ice-cold water will chill a bottle, 20 minutes will make it quite cold, and 45 minutes will make it ice-cold. Salt may be added to the ice water to chill the bottle even faster.

Basic wine service "how to's" are detailed in Chapters Seven and Eleven. Any dramatic flourishes in wine service should be professional. In an informal atmosphere, perhaps servers can be a bit more daring. Showmanship should always stop short of damaging the wine through overchilling, overheating, overshaking, or overhandling the wine. The following general rules will be useful:

♦ Buckets and stands can be used for all wines. Ice is used only for wines that have to be chilled, generally white, rosé, Champagnes, and sparkling wines.

♦ Decanting cradles can be used for "show" with all red wines when the wines have no sediment and are placed on the table or served from a sidestand. However, if decanting cradles are used to decant wines with sediment, they should be used for decanting those wines only. Other red wines that do not require decanting should then be stood on the table or sidestand.

♦ All wines can be decanted into crystal carafes in front of the customer for show. Only rare wines and/or wines with sediment really have to be decanted.

♦ The sommelier, wine steward, or server can taste a few drops of the wine from a silver tastevin if that is the policy of the operation. Generally, unless there is an official sommelier, this practice is outdated and the guest tastes his or her own wine. Unless it is refused by the guest, members of the service team do not generally taste guests' wines.

♦ If the operation has an elaborate lever-action corkscrew, it can be used instead of the professional waiter's corkscrew.

♦ Champagne corks can be popped instead of slowly eased from the bottle if for a gala celebration (e.g., New Year's Eve at midnight, when decorum and proper service are thrown to the wind). The proper way to open a bottle of sparkling wine is to ease the cork from the bottle with a whisper, without a pop, resulting in more wine to drink and less foaming out of the bottle.

♦ If the mushroom cap of the cork on a bottle of sparkling wine should break off, the bottle must be opened with a regular corkscrew. Unfortunately, this is a difficult procedure as such corks are highly compressed and of excellent quality so it will be difficult to get the worm into the cork. Also, the pressure of the wine will prevent a controlled release of the cork. This is a dangerous procedure and should only be done in the service bar area or in the kitchen, not in the dining room or a public space area, and certainly not in front of any guests.

Opening Old Port

Ports and other wines that are destined for a long life are sealed with extra-long corks that fill the neck of the bottle and mushroom slightly when they enter its shoulders. If the bottle has been recorked within the past 15 or 20 years, there should be little problem in removing it. Long-lived bottles are periodically recorked because the wine outlives the cork. If the cork is very old, it may be impossible to remove, even using a tweezer-type extractor, because the cork may crumble into the bottle.

In this instance, the best course is to knock the neck off the bottle. Hobby stores sell a "glass saw" for making artifacts from bottles, which works well to score the bottle so that a sharp rap with a heavy knife breaks the neck cleanly. A conventional glass cutter also works, or the bottle may be scored, with less consistent success, with a heavy knife. When Port was in vogue in the old homes of England, special tongs were heated to "red-hot" in the fireplace and applied to the bottle's neck to cause the glass to snap. These are available today from wine connoisseur mail-order houses, and are also great conversation items.

Glassware

Type of Glass

The numerous specialty glasses for wine should be evaluated against the ideal glass. The ideal glass is thin clear crystal without flaws or any design, etching, or hint of color or tint. It has a sturdy base and comfortable stem. The bowl is slightly tulip-shaped with an opening that will comfortably admit a nose. The glass will hold 8 ounces filled to a brim unmarred by a safety rim. This allows for a comfortable 3- to 4-ounce pour and sufficient room in the glass to swirl the wine, releasing the bouquet.

Colored, etched, or textured glasses distort a wine's appearance. Clumsy glasses—for example, little bowls on high stems—make it difficult to examine the wine. Thick glasses give it a different "mouth-feel." Diminutive glasses do not concentrate the bouquet, and oversize glasses can cause the delicate bouquet of the old wines usually served in this kind of glass to disintegrate from exposure. (Figure 11.3 on page 214 shows traditional shapes of wineglasses.)

Cleanliness of Glassware

Any smell or taste adhering to glassware will be communicated to the wine and amplified by it. Specialty glassware is often stored in closed cupboards, where it acquires a musty odor. Many modern detergent compounds leave a slight scent or a chemical residue to prevent streaking that can be communicated to the wine. It is best to redip washed wineglasses in hot water, dry and polish them with lint-free clean cloths prior to use to ensure perfect cleanliness.

Wine and the Menu

Wine may be enjoyed any time of the day: a glass of wine instead of a cup of coffee late in the afternoon, wine coolers instead of sodas, wine as a cocktail, wine as an evening refreshment, and wine as a nightcap. Wine is also the perfect accompaniment to meals since a wine can be found to complement most foods. Wine should not be just a special-occasion treat.

Wine and Food Combinations

Tradition, based on others' trial-and-error "research" of various wine and food combinations, can serve as a reasonable guide for recommending the "right" wine to have with various foods.

It is a mistake to dismiss tradition; it is also a mistake to revere it. Tastes differ. For example, Europeans generally like wines warmer than do Americans; many Germans enjoy their white wines as much as 10 degrees warmer. Tastes change. Heavy Port is not drunk today as often as yesteryear, but it still dominates much of traditional English wine and gastronomic literature.

The Color Question

The traditional rule, red wine with red meat, white wine with white meat, is a good beginning point. However, the individual wine and the individual dish must be considered.

White wines, being for the most part lighter than red wines, generally suit lighter dishes such as fish, fowl, veal with light sauces. Red wines generally complement heartier dishes, grilled foods, foods with heavy or spicy sauces.

Red wines taste less pleasant when served with seafood dishes that have a taste of the sea or a metallic alkaline taste, especially when the red wine itself is rough and full of unmatured tannins. On the other hand, quality reds seem to suit the "meatier" and fattier fishes like salmon very well. A convenient resolution of the problem can often be found in traditional classic cookbooks: If there are dishes for a fish using red wine, then red wine is a suitable accompaniment.

Another age-old "rule," that rosé wine can be drunk with everything, is vulnerable. A very light rosé is really a white wine, whereas a very dark rosé resembles a light red. Even when the rosé is truly pink, it may be dry or sweet, full-bodied or light, or have any number of other qualities that will make it unsuitable for certain dishes.

Wine "Rules"

A great deal of the disenchantment with traditional wine rules is related to some arbitrary standards that European (mostly English) connoisseurs developed ages ago. In a time and place when wines were available and inexpensive, it was possible to decide to drink specific wines with specific foods on the basis of many arbitrary notions. The geographic basis has some merit; two such rules seem to have become part of traditional ritual: (1) Drink national wines with national dishes, and (2) drink regional wines with food produced in the same region.

Many people enjoy *paella,* a Spanish dish of rice, vegetables, saffron, and other seasonings garnished with chicken, seafood, and sausage. Served with a fine, robust Spanish wine, such as a Rioja, it's terrific. However, it would be basically unfair to the dish to serve it with a light-bodied wine just because the wine happened to be Spanish.

Drinking regional wines with regional foods has a certain romantic appeal and practical basis. Imagine eating a local goat cheese with a glass of local wine while the goat from whose milk the cheese was made munches on the wine-producing grapevine. Complementary flavors will be carried over in both the wine and the cheese. The romance may be sufficient compensation for mediocre cheese or mediocre wine if you are visiting that region, but when you are several thousand miles away, practicality displaces romance. Some regions that produce excellent food products (e.g., cheese or fruit) produce mediocre wine, if they produce any wine at all. Normandy, France, for example, produces Brie, Camembert, and other excellent cheeses, apple cider, and perry. Generally, the wine connoisseur would opt for French Burgundy or Bordeaux rather than apple juice or perry as a suitable drink for cheese. By the same token, fine wine is produced from grapes that grow in inhospitable places on unfertile soils, so the purchaser must use judgment in matching regional foods and wine.

Water and Wine

Traditional wine connoisseurs are repelled at the idea of watering wine, pouring wine over ice cubes or cracked ice, or in any way adulterating it. Without doubt, adding water fundamentally changes a wine. The drinker must really ask why the particular bottle was purchased: Will adulteration with water, carbonated water, fruit juice, or a soda drink diminish or enhance the wine's "purchase points"? If a wine was purchased because it promised to be refreshing, its refreshing qualities may be enhanced by adding soda and fruit, or cooling it (and diluting it) over ice.

The same logic can apply to wine that accompanies meals. It may be more satisfying to drink great draughts of a wine cooler or jug wine with a spicy spaghetti dish than to sip decorously at a glass of fine aged wine. In this circumstance, conversion of a jug wine into a wine cooler or a wine-with-ice drink would seem reasonable, just as drinking a vin ordinaire rather than a special vintage wine would suffice.

Basics of Matching Food and Wine

It is important to know the traditional rules (sprinkled with elements of common sense) regarding the matching of foods with wines, even if those rules will be ignored because of some transcendent personal taste:

1. For the full value of the wines to be appreciated when several wines are served during a meal, they must be ordered so that the palate is not prejudiced by the first wines drunk:

- Dry before sweet.
- White before red, unless the serving order violates the dry-before-sweet rule.
- Young before old.
- Light before full-bodied, unless the serving order violates the dry-before-sweet rule.
- Modest before great.
- Table wines before fortified wines.

2. If Champagne is used as a table wine, its service should be continued throughout the meal.
3. Dishes that contain vinegar—or foods that contain it—do not go well with wine: pickles, mustard, ketchup, salad dressing, and so on.
4. Madeira or Sherry is preferred with soup, unless the soup contains wine, in which case no wine is served.
5. Once a person has developed a palate for fine wines, a few very good wines are more enjoyable than a plethora of jug wines or inexpensive wines.
6. Delicate wines should be served with delicate dishes.
7. Robust foods need robust wines to "stand up" to them.
8. Sweet wines should be served with sweet fruits; no wine with citrus fruits.
9. Very sweet wines are used for dessert courses.
10. Either Champagne or no wine should be served with egg dishes.

Guidelines for Personal Matches of Wine and Food

The wine and the dish should be of about the same flavor order; for example, simple and direct (pizza with a California mountain red) or complex but subtle (Pike Quenelles in Nantua Sauce with a Montrachet). A few guidelines for matching wines and food are shown in the chart at the end of this chapter.

The following guidelines will be useful:

♦ The wine should either harmonize with the food and complement it, or contrast with the food and thereby reveal it. For example, a dry white wine harmonizes with seafood, and a sweet white wine contrasts with curry.

♦ Avoid serving wines with foods having qualities that, if found in the wine, would fault it. Vinegary foods and wine are the best example; a taste of vinegar in a wine is a great defect; therefore, avoid wine with salads with vinegar dressings. Also, eggs do not go well with wine because of their basic sulfury taste; the blue cheeses go better with fortified wines than with table wines because fortified wines have sufficient alcohol to support the taste of strong cheeses.

♦ Consider the preparation of foods, not just the main ingredients. Ideally, a wine served with barbecued chicken would be different from the wine served with roasted chicken. In the first case, the wine has to stand up to ketchup, garlic, chili powder, lemon juice, and pepper.

♦ Consider the weather, the mood of the party, and the occasion. For example, a great full-bodied wine is not the best choice on a hot, humid day, no matter what the menu. Likewise, fortified wines in close, heated, crowded dining rooms make the diners feel uncomfortable.

♦ A good dish merits a good wine even if it is a "wrong" wine. In other words, it is better to serve a quality red wine with a fish than to serve an indifferent white wine just because it is white.

WINE AND FOOD SUGGESTIONS

Food	Wine
Beef	Dry, full-bodied reds, e.g., Bordeaux, Cabernet Sauvignon*
Cake	Cold, sweet white wines
Cheese	
Gruyère (Swiss)	Dry white
Edam or Gouda	Full-bodied red
Semisoft Port Salut	All colors, dry and fruity
Blue	Fortified
Soft	Light reds
Fresh	Crisp whites, Rosés
Processed	All colors, dry
Goat and sheep	Tart whites
Chicken	
Hot	Dry, red or white wines*
Cold	Dry or semidry whites
Spicy	Gewürztraminer
Coffee	Vintage Port
Desserts in general	Sweet whites, Sauterne, Champagne, Asti Spumante
Eggs in general	No wine
Omelettes	Light reds, Champagne

Food	*Wine*
Fish	Dry delicate-to-full-bodied whites, Rosés*
Foie Gras	
Preserved Fresh	Fortified dry reds
Fruit	
Apples	Red wines
Citrus fruits	No wine
Strawberries	Port, Champagne, sparkling wines
Other fruit	Sweet whites
Game	Full-bodied red, e.g., Burgundies, Pinot Noir*
Ham	Light white, Beaujolais*
Ice cream	No wine
Salad	No wine
Lamb	Light to full-bodied red, e.g., Burgundy, Pinot Noir*
Poultry	Light red, e.g., Beaujolais*
Shellfish	
Raw	Dry white, e.g., Chablis
Cooked	Delicate-to-full-bodied white, e.g., Chablis, Chardonnay*
Smoked foods	Dry white, dry fortified, e.g., Gewürztraminer
Soup	Dry fortified
Made with wine	No wine
Veal	Light to full-bodied reds, whites, e.g., Burgundy, Pinot Noir, Chardonnay*
Vegetables	Semidry whites, Rosés

*Choice should depend on the preparation of the main ingredient and key sauce ingredients. For instance, with a tomato-based sauce usually a red wine, e.g., Beaujolais, Burgundy, Bordeaux, Cabernet Sauvignon, Rioja, Chianti, Pinot Noir, Zinfandel, could be served. If a cream-based or vegetable-puree-based sauce is used with the main dish, perhaps a full-bodied, smooth white wine such as a Chardonnay could be served. With a broth-based entrée or one with a very light sauce, a very light wine would most likely complement the food.

Match the richness of the food with the richness of the wine. Match spicy dishes with a robust wine, a fruity wine, or a spicy wine, e.g., Gewürztraminer (white) or Zinfandel (red).

Wines of the World

❖ ## WINES OF THE UNITED STATES

Many countries are producing great wines: France, Germany, Italy, the United States, Spain, Australia, just to name a few. Many other countries produce pleasant, even excellent, wines that should also be tried.

The United States is a relative newcomer to the production of great wines since it was necessary to rebuild the American wine industry after the repeal of Prohibition. The quality wine production, in general, is centered along the West Coast, primarily the cooler coastal regions of California, Oregon, and Washington, while the jug wine production is centered in the warmer microclimates of central California.

The grape varieties from which great European wines are made—Cabernet Sauvignon, Chardonnay, Pinot Noir, Riesling—prosper in several microclimates of the United States. In California, the largest of the 20 or more wine-producing states, much sun and fertile land result in great harvests.

In addition to fine wines, the United States, especially California, produces many inexpensive and medium-priced table wines that offer extremely good value. The American understanding of agriculture as a business, utilizing mass production techniques with quality control, mechanized packaging, and mass distribution, effect tremendous economies that at least in part are transmitted to the consumer. American jug wines, sold in gallons, with labeling that emphasizes the brand name of the producer, are generally better values than imported inexpensive wines. American bottled wines also provide medium-quality wines at moderate prices. They offer individuality and personality, often identified by the growing region and varietal name (e.g., Napa Valley Zinfandel) or, as in France, by the name of the vineyard and the varietal name.

The law in the United States is rigid and controlled so as to protect consumers, but it is not consistent with the wine statutes of the European producers on the Common Market. Names of foreign wines and foreign wine regions are not protected. *Sherry*, for example, must qualify by alcohol content, nothing else. *Sauternes* and *Chablis* are specific, individual, distinct French grape-growing and wine-producing regions, but in the United States, Sauterne and Chablis can be applied to any wine. It is legally possible in the United States to draw wine from the same vat to be marketed as both Sauterne and Chablis.

Some of the more expensive American wines are labeled with a name they are known for: the vineyard, a reputable vintner, a growing region, and/or a grape variety. The less expensive American wines are often labeled to emphasize a brand name, which like that of most American products is meant to provide assurance of consistent quality irrespective of season, harvest, or growing conditions of the raw materials.

Table Wine Varieties

Three types of wine are produced:

1. Varietal wines named after specific grape varieties (e.g., Chardonnay, Pinot Noir).
2. Generic wines named after European wine-growing regions (e.g., Burgundy).
3. Proprietary or brand-name wines produced to conform with a commercial standard (e.g., "Opus One" or Thunderbird).

Varietal names are further qualified by the name of the vintner and the growing region. The better red varietals include Cabernet Sauvignon, Pinot Noir, Gamay, Barbera, and Zinfandel. Notable white varietals are Chardonnay, Riesling, Sauvignon Blanc, Pinot Blanc, Chenin Blanc, and Traminer. Rosé varietals are best represented by Grenache, Grignolino, and Gamay.

Generic wines are named for the most famous European wines, based on the region the grapes are grown in: Burgundy, Chablis, Sauterne, Tokay, Chianti, Port, Claret, Beaujolais, Tavel, Champagne, and sparkling Burgundy. The quality of the product varies enormously, since there is no control over the application of these names to California wines.

Proprietary or brand-name wines are merchandised by both small and large concerns attempting to offer a marketable, accessible, consistent product. The wine may be a table wine or a refreshment wine bolstered with fruit juices and spices.

Federal Laws

Federal laws differentiate between the grape varieties used in the wines, but otherwise are fairly standard.

♦ If a wine is labeled as a *Vitis vinifera* varietal (e.g., Chardonnay, Pinot Noir, Cabernet Sauvignon, Sauvignon Blanc, Pinot Blanc, Semillion, Chenin Blanc, Gewürztraminer), then 75 percent of the grapes used to make that wine must be the grape stated on the label.

♦ However, if a wine is labeled as a *Vitis labruscana* varietal (a native American variety such as Catawba, Delaware, Dutchess, Diamond, Elvira, Concord), only 51 percent of the grapes used to make that wine must be the grape stated on the label. This is due to the strong flavor of native grapes.

♦ If a specific geographic area is specified on the label, (e.g., Napa Valley), then 75 percent of the grape juice used to make the wine must be from that area.

♦ If a specific viticultural area or microclimate area is specified on the label, such as Caymus, Oakville, Alexander Valley, or Carneros, then 85 percent of the grape juice used to make the wine must be from that area.

♦ If a specific viticultural area and a vintage year are stated on the label (e.g., Diamond Creek Vineyards, Red Rock Terrace 1987), then 95 percent of the grape juice used to make that wine must come from the specified area and the specified year.

♦ If a label states "Estate bottled," then 100 percent of the grapes used to make that wine must have been grown on the winery's premises.

♦ If a label states "Produced and Bottled by . . . ," then 75 percent of the grapes used for that wine must have been crushed and fermented at that winery, but not necessarily grown there.

West Coast Vineyards

Although some wine grapes are produced in most counties in California and sold to large vintners, there are three major production areas:

1. The north Coast Range near San Francisco, as well as vineyards in Oregon, Washington and Utah, which are noted for the best wines being produced in the United States.
2. The Central Valley, known for jug and bulk wines as well as some sherries and sweet wines.

3. The inland, hot Cucamonga district of Southern California, which produces fortified wines.

New York State Wines

The production of New York State (and the rest of the United States) is minimal, a few million gallons of fairly indifferent wine, distinguished only by the particular taste of the native American grape species (Labrusca grape varieties) and their hybrids. Often these grapes, being strong in flavor, produce wines described as "foxy" or "grapey" or "wild."

The main Eastern growing region is the Finger Lakes area in the southwestern part of New York, which in topography and exposure somewhat resembles the growing regions of the Rhine.

Numerous varietals are produced including Delaware, Catawba, Elvira, Isabella, Salem, Vergennes, and Riesling. The region also produces an assortment of fortified wines, Sherries, Ports, Muscatels, and sweet dessert wines such as Tokay and sweet Catawba. In addition, sacramental and sparkling wines are produced.

The region's best efforts include such German derivative wines as Spätlese, Beerenauslese, and Trockenbeerenauslese. These might be tried before their prototypes as they are certainly representative of the type and less expensive than the German wines.

❖ WINES OF FRANCE

France is generally regarded as the primary place of origin of fine quality wines. While other countries are now producing quality wines, France has long been noted for refined, smooth, complex wines derived from certain defined and regulated grapes, making those wines the standard against which winemakers from any other countries must compare their wines.

Because, historically, French wines set the standard, this section extensively reviews the wines of France to provide a "base reference" for comparing wines of other countries. While Robert Mondavi, say some, produces the standard for a quality, consistent jug wine, most connoisseurs still regard the wines of France as the epitome of fine wines.

The intent of this section is to provide not only an understanding of French wines but also an insight into why they are considered wines to emulate.

France produces tremendous quantities of wine, imports tremendous quantities of wine, and consumes tremendous quantities of wine. Few of these million of gallons have any interest for buyers outside France, as they are *vin ordinaire* or table wine.

About 10 percent of the wine produced is *Appellation Contrôlée* (A.O.C.), entitled to an official guarantee of origin and quality; of these, a much smaller percentage is likely to have real merit, and a still smaller number has the quality of appeal to justify its exportation. For example, the French *département* (state) where Bordeaux is located, the *Gironde,* which ranks fourth in total gallonage production but first in production of Appellation Contrôlée wines (75 percent), has over 50,000 vineyard proprietors, and only 2,000 of them are noteworthy. Of these, fewer than 200 have produced world-renowned Bordeaux wines. Many purchasers are aware that if a wine is French (Bordeaux, Burgundy, Côtes-du-Rhône, etc.) it might be great, but its being French is not necessarily a guarantee that it is good.

Most French wines are classified, named, controlled, and labeled on the basis of a very precise geographic organization of the four million acres of vineyards. In some respects, this organization follows the political divisions of the country. Control and quality standards increase as smaller and smaller areas are addressed. For example, a very ordinary wine from over 250,000 acres of vineyards in Gironde can be called a Bordeaux, but only a wine of definite quality standards from a specific 110 acres can be called a *Château Latour.* It is produced in the *Pauillac* commune of the *Haut-Médoc* area of the *Médoc* region in the département of Gironde, France.

Regional names may or may not be Appellation Contrôlée names themselves. Bordeaux is, but Loire is not. This means that a wine called Bordeaux on the label is an Appellation Contrôlée wine. A wine from the Loire will either be called by an Appellation Contrôlée name referring to a smaller area within the Loire: for example *Anjou Appellation Contrôlée;* the name of a smaller region designated V.D.Q.S. (*Vins Délimités de Qualité Supérieure*), which is a geographic classification system of wines that are less notable than the Appellation Contrôlée; or any name whatsoever so long as it is not fraudulent and does not infringe on the geographic classifications.

Bordeaux

Comprising about 110,000 areas of vineyards, the Bordeaux region produces wines that are among the best in the world. As a producing area, Bordeaux is also unrivaled for variety. Three distinct types of fine wines are produced: (1) the Bordeaux red wines (clarets), (2) the white wines (e.g., those of Graves), and (3) the sweet white wines (e.g., those of Sauternes). In each of these categories, the best Bordeaux wine equals or exceeds the best wine of regions devoted exclusively to the production of a single type. In addition to the fine wines, Bordeaux produces "Bordeaux" wines that are vin ordinaire, as well as *Bordeaux Supérieur* which is slightly better. These are consumed locally as table wine.

The Bordeaux region is divided into several districts including: *Médoc, Graves, Sauternes, Saint-Émilion, Pomerol, Côtes de Bourg, Entre-deux-Mers* (see the box on p. 298 for a complete list). Each of these is divided into further subdivisions, or parishes, which contain individual vineyards. The great wines of Bordeaux take their names from vineyard names: *Château Margaux, Château Latour, Château Haut-Brion,* and so on.

The leading châteaux of four districts have been classified on the basis of the consistent quality of their wines: Médoc, Sauternes, and Saint-Émilion in 1855, and Graves in 1953. Based on the rating received by each vineyard in those districts, the wine was "merited."

Although there is certainly injustice (and some obsolescence) in the following charts, they are widely and justifiably approved. However, users of the charts should not over conclude from these rankings. The lists include only the absolutely greatest Bordeaux wines, not all of them. There remain excellent châteaux classified as *Crus Exceptionnels, Crus Bourgeois Supérieurs,* and *Crus Bourgeois* all below the fifth growth wines (*Cinquièmes Crus*). There are also excellent châteaux, some the equal of the classed growths, that have not been classified and fine Appellation Contrôlée wines that bear parish names. Within the lists, the ranking is only relative. For example, the difference between the wines is very, very small in the Médoc list, which ranked the 15 best red Bordeaux, although the first growths command higher prices than the second growths, and within growths the higher-ranking wines (the first wine of the second growth) are more expensive.

Generally speaking, the great Bordeaux wines have either (1) the name of a district (e.g., Médoc, Graves), (2) the name of a specific village or a *première* district (e.g., Pauillac, Margaux, St. Julien), or (3) the actual name of the château where the wine was made (e.g., Lafite-Rothschild). As a rule of thumb, the more specific the name on the label, the better the quality of wine.

If the wine was produced and bottled at the same estate, the label will state and, with the better wines, the cork will be stamped with these words: *Mis en bouteilles au château*. Usually, this is also an indicator of quality.

1855 CLASSIFICATION OF MÉDOC RED WINES

Château	*Commune*
First Growths	
Lafite-Rothschild	Pauillac
Latour	Pauillac
Margaux	Margaux
Haut-Brion	Pessac (Graves)
Second Growths	
Mouton-Rothschild*	Pauillac
Lascombes	Margaux
Rausan-Ségla	Margaux
Rauzan-Gassies	Margaux
Léoville-Lascases Las-Cases	Saint-Julien
Léoville-Poyferré	Saint-Julien
Léoville-Barton	Saint-Julien
Durfort-Vivens	Margaux
Gruaud Larose	Saint-Julien
Brane-Cantenac	Cantenac-Margaux
Pichon-Longueville	Pauillac
Pichon-Longueville (Comtesse de Lalande)	Pauillac
Ducru-Beaucaillou	Saint-Julien
Cos d'Estournel	Saint-Estèphe
Montrose	Saint-Estèphe
Third Growths	
Giscours	Labarde-Margaux
Kirwan	Cantenac-Margaux
d'Issan	Cantenac-Margaux
Lagrange	Saint-Julien
Langoa-Barton	Saint-Julien
Malescot-Saint-Exupéry	Margaux
Cantenac-Brown	Cantenac-Margaux
Palmer	Cantenac-Margaux
La Lagune	Ludon
Desmirail	Margaux
Calon-Ségur	Saint-Estèphe
Ferrière	Margaux
Marquis d'Alesme-Becker	Margaux
Boyd-Cantenac	Margaux
Fourth Growths	
Le Prieure-Lichine	Cantenac-Margaux

Château	Commune
Saint Pierre	Saint-Julien
Branaire-Duluc	Saint-Julien
Talbot	Saint-Julien
Duhart-Milon	Pauillac
Pouget	Cantenac-Margaux
La Tour-Carnet	Saint-Laurent
Lafon-Rochet	Saint-Estèphe
Beychevelle	Saint-Julien
Marquis-de-Terme	Margaux
Fifth Growths	
Pontet-Canet	Pauillac
Batailley	Pauillac
Grand-Puy-Lacoste	Pauillac
Grand-Puy-Ducasse	Pauillac
Haut-Batailley	Pauillac
Lynch-Bages	Pauillac
Lynch-Moussas	Pauillac
Dauzac	Labarde-Margaux
Mouton-Baron-Philippe	Pauillac
du Tertre	Arsac-Margaux
Haut-Bages-Libéral	Pauillac
Pédesclaux	Pauillac
Belgrave	Saint-Laurent
Camensac	Saint-Laurent
Cos-Labory	Saint-Laurent
Clerc-Milon-Mondon	Pauillac
Croizet Bages	Pauillac
Cantermerle	Macau
Exceptional Growths	
Villegorge	Avesan
Angludet	Cantenac-Margaux
Chasse-Spleen	Moulis
Poujeaux-Theil	Moulis
La Couronne	Pauillac
Moulin-Riche	Saint-Julien
Bel-Air-Marquis d'Aligre	Soussans-Margaux

*Decreed a First Growth in 1973.

1855 CLASSIFICATION OF SAUTERNES WHITE WINES

Château	*Commune*
First Great Growth	
d'Yquem	Sauternes
Guiraud	Sauternes
La Tour-Blanche	Bommes
Lafaurie-Peyraguey	Bommes
Haut-Peyraguey	Bommes
Rayne-Vigneau	Bommes
Rabaud-Promis	Bommes
Sigalas-Rabaud	Bommes
Coutet	Barsac
Climens	Barsac
Suduiraut	Preignac
Rieussec	Fargues
Second Growths	
d'Arche	Sauternes
Filhot	Sauternes
Lamothe	Sauternes
Myrat	Barsac
Doisy-Védrines	Barsac
Doisy-Daëne	Barsac
Suau	Barsac
Broustet	Barsac
Caillou	Barsac
Nairac	Barsac
de Malle	Preignac
Romer	Fargues

1953 CLASSIFICATION OF
GRAVES RED AND WHITE WINES

Château	*Commune*
Classed Red Wines	
Haut-Brion	Pessac
Bouscaut	Cadaujac
Carbonnieux	Léognan
Domaine de Chevalier	Léognan
de Fieuzal	Léognan
Haut-Bailly	Léognan
La Mission-Haut Brion	Pessac
La Tour Haut-Brion	Talence
La Tour-Martillac	Martillac
Malartic-Lagravière	Léognan
Olivier	Léognan
Pape-Clément	Pessac
Smith-Haut-Lafitte	Martillac
Classed White Wines	
Bouscaut	Cadaujac
Carbonnieux	Léognan
Domaine de Chevalier	Léognan
Couhins	Villenave d'Ornon
Haut Brion★★	Pessac
La Tour-Martillac★	Martillac
Laville-Haut-Brion	Talealence
Malartic-Lagravière★	Léognan
Olivier	Léognan

★Added in 1959.
★★Added in 1960.

1955 CLASSIFICATION OF SAINT-ÉMILION WINES

There is no importance (distinguishing characteristics) attached to the communes of Saint-Émilion wines. Since the châteaux on this list are protected by the A.O.C. laws, they have the weight of an Appellation Contrôlée, while those of Médoc do not. Names are prefaced by Château unless otherwise indicated.

First Great Growths

Ausone
Cheval Blanc
Beauséjour-Dufau-Lagarrosse
Beauséjour-Fagouet
Bélair
La Fourtet

Canon
Figeac
La Gaffelière
Magdelaine
Pavie
Trottevieille

Great Growths

Angelus
L'Arrosée
Balestard-la-Tonnelle
Bellevue
Bergat
Cadet-Bon
Cadet-Piola
Canon-la-Gaffelière
Cap de Mourlin
Chapelle-Madeleine
Chauvin
Clos des Jacobins
Clos de Madeleine
Clos Saint-Martin
Corbin
Corbin-Michotte
Coutet
Croque-Michotte
Curé-Bon
Fonplégade
Fonroque
Franc-Mayne
Grand-Barrail-Lamarzelle-Figeac
Grand-Corbin
Grand-Mayne
Grandes Murailles
Grand Pontet
Gaudet-Saint-Julien
Jean Faure
La Carte

La Clotte
La Cluzière
La Couspaude
La Dominique
Lamarzelle
Larcis-Ducasse
Larmande
Laroze
Lasserre
La Tour-Figeac
La Tour-du-Pin-Fingeac
Le Couvent
Le Prieuré
Mauvezin
Moulin-du-Cadet
Pavie-Décesse
Pavie-Macquin
Pavillon-Cadet
Petit-Faurie-de-Soutard
Ripeau
Saint-Georges-Côte-Pavie
Sansonnet
Soutard
Tertre-Daugay
Trimoulet
Trois-Moulins
Troplong-Mondot
Villemaurine
Yon Figeac

MAJOR APPELLATIONS CONTRÔLÉES FROM BORDEAUX IN ADDITION TO BORDEAUX A.O.C.

White

Graves	Sainte-Croix-du-Mont
Cérons	Loupiac
Sauternes	Sainte-Foy-Bordeaux
Barsac	Graves de Vayres
Entre-deux-Mers Première	Côtes de Bordeaux
Côtes de Bordeaux	Saint-Macaire

Red

Médoc	Blaye
Haut-Médoc	Graves
Saint-Estèphe	Lussac-Saint-Émilion
Pauillac	Montagne-Saint-Émilion
Saint-Julien	Parsac-Saint-Émilion
Margaux	Puisseguin-Saint-Émilion
Moulis	Saint-Georges-Saint-Émilion
Listrac	Côtes de Castillon
Pomerol	Lalande de Pomerol
Côtes de Fronsac	
Bourg	

Médoc

The Médoc region produces most of the best-known Bordeaux red wines, famed for their robustness, complexity, smoothness, and full body. Médoc is sometimes further "divided" into two subregions: *Haut-Médoc* and Médoc or *Bas-Médoc.*

Haut-Médoc wines are generally of high quality, varying from the full-bodied, somewhat fruity Saint-Estèphe wines to the somewhat lighter and softer Margaux, with Saint-Julien in between. Pauillac offers some of the great big yet mellow wines of France such as *Château Lafite* and *Château Latour.* The major grape used is the Cabernet Sauvignon, which often takes several years to mellow due to the tannins and complexity of flavors. Often, the wines are "softened" with Merlot grapes to make a very smooth wine, ready to drink sooner.

Graves

Graves produces both red and white wines but is better known for its white wines, varying from France's finest full-bodied dry white wines to its sweet dessert wines (Sauternes/Barsac and Cérons). Its red wines can be excellent, and at their very best more expensive than the whites. They are not as delicate as Médoc reds, but they have remarkable clarity, good body, and bouquet. *Château Haut-Brion*, although a Graves, was classed in 1855 with the Médoc. A reclassification today might well include other red Graves as the peers of some of the great Médoc wines.

The white wines of Graves vary considerably from the dry Haut-Brion and Couhins, for example, to the sweet and luscious Sauternes, and also include *Barsacs* and *Cérons*, which moderate between the typical dry Graves and the sweet Sauternes.

Sauternes

Sauternes, a part of the Graves region, produces unique white wine from five communes: *Barsac*, sold under its own name, and *Sauterne*, *Bommes*, *Fargues*, and *Preignac*, sold as Sauternes. Although the region also produces some dry Sauterne (genuinely dry, not simply less sweet) and some red wine, only the golden, sweet, high-alcohol dessert white wine can be called Sauternes A.O.C. Red wines are rare from Sauternes.

Barsac wines tend to be a little lighter and drier than Sauternes, and perhaps owe more of their bouquet to the perfume of maturation than to the aroma of the *Sémillon*, *Sauvignon Blanc*, and *Muscadelle* grapes, from which they (like all Bordeaux white wines) are made.

Like the great German *Trockenbeerenauslese*, Sauternes are made from grapes individually hand-picked when they are literally overripe and covered with noble rot (*Botrytis cinerea*).

The term *Haut-Suaterne* has no official meaning; it is used in the trade to designate a very sweet Sauterne.

Saint-Émilion

Saint-Émilion often produces a greater quantity of excellent red wines than Médoc. They are usually higher in alcohol, darker, and more full-bodied. At their most typical, they seem to moderate between the classic Bordeaux wines and the Burgundies.

Although they are classed with other châteaux as First Great Growths, the *Château Ausone* and the *Château Cheval-Blanc* are outstanding. Both would rank with the best Médoc wines.

Pomerol

The crimson, full-bodied, velvety wines of Pomerol, adjoining Saint-Émilion, moderate between it and the Médocs. They have much of the delicacy and mellowness of the Médocs. Pomerol contains only 1,500 acres, compared with the 16,000 of Saint-Émilion, and except for a few great châteaux (*Pétrus, Certan-Giraud, Certan-de-May, Vieux-Chateau-Certan, La Conseillante,* and *Trotanoy*) and some really excellent wine sold as Pomerol, it is not well represented.

Other Bordeaux Regions

Fronsac, next to Pomerol, produces red wines that resemble Pomerols, although they lack the delicacy of a Pomerol. Bourg and Blaye also produce quality red wines more in the Saint-Émilion family than the Médoc.

Entre-deux-Mers covers nearly 20 percent of the Bordeaux region. It produces large quantities of red and white wine. Quality and characteristics vary tremendously because of the size of the area. It includes numerous Appellations Contrôlées that have their own small reputations: Premières Côtes de Bordeaux for good white wines; Graves de Vayres for Graves-type wines; Loupiac for Sauterne-type whites; and Sainte-Croix-du-Mont for Barsac-type whites.

The wine sold as Entre-deux-Mers A.O.C. is white; the red wine is sold as Bordeaux A.O.C.

Burgundy

While the system of wine classification and nomenclature in Bordeaux is complex, it is orderly. A name on the label of Bordeaux wine has a precise referent. Although obviously not every bottle called Pauillac A.O.C. will be like every other bottle, definite standards have been met by all bottles bearing the name, and the name itself is not misleading. A wine further defined as Château Latour is exactly that wine, no other, identical in all respects to every other bottle of that same year.

In Burgundy, the system of wine classification is not only complex but also confusing. Problems of nomenclature, resulting in a profusion of different wines or a puzzling difference in quality between bottles with the same name on the bottle, have four causes:

1. The vineyards in Burgundy are small, varying from minuscule to modest, and small proprietors of different vineyards sell their wines under the names of their villages, not their own names.
2. Some vineyards have numerous owners, each of whom independently produces wine but legitimately and confusingly uses the name of the vineyard.

3. The villages of the Burgundy region have taken the name of the most famous vineyard in the region and attached it to their own.
4. Some vineyards also have the right to attach the name of the most famous vineyard to their own.

These practices can make evaluation of Burgundy wine based on past experience with them quite difficult. They also make a Bordeaux-style classification system impossible. For example, the wine that bears the name *Chambertin* (and no other qualifiers) is among the most famous and excellent Burgundies. But which Chambertin? The vineyard has 32 acres owned by 25 people, each of whom produces Chambertin. Even if the winemakers were all equally skilled, and they are not, each small section of vineyard has its own microclimate. Thus, there are at least 25 Chambertins. There may be even more, as shippers can buy several Chambertins and make their own. In addition, because of the reputation of the wine, the nearest village, Gevrey, is called *Gevrey-Chambertin*, and there may be as many as 10 or 12 vineyards (there could be any number) who use the village name as their name. Finally, there are vineyards in the locality that have been given the right by the A.O.C. laws to attach the name Chambertin to their own names: Chambertin-Clos de Bèze, Ruchottes-Chambertin, Chapelle-Chambertin, Mazoyères-Chambertin, and others.

Short of learning about the wine by tasting it, generally the best wine is the one that bears the name of a famous vineyard like Chambertin without qualification, shipped by a well-known shipper. Next best are those that bear the vineyard name, followed by the wines that bear hyphenated vineyard names, then those that bear village names plus the words premier cru, and finally those that bear only village names. The ranking of those bearing village names, the most frequent problem, must be on the basis of the reputation of the shipper or importer.

Major Growing Regions of Burgundy

The Burgundy production area is a narrow strip that extends about 130 miles. Until the nineteenth century, the entire area was cultivated, but after the *Phylloxera* (vine-attacking plant lice) destroyed most French vineyards, only the best were replanted.

The Basse-Bourgogne comprises of the Côte d'Or, Côte Chalonnaise, Maconnais, and Beaujolais growing areas.

The Côte d'Or, which produces the classic Burgundies, is further divided into the Côte de Dijon, the Côte de Nuits, and the Côte de Beaune. The production of the Côte de Dijon is limited and not extraordinary. The Côte de Nuit produces the most famous red Burgundies (with the exception of Beaujolais), while the Côte de Beaune produces the most famous white Burgundies (with the exception of the Chablis, produced in the Yonne area, Haute-Bourgogne).

In essence, the Burgundies, as they are seen in the United States, consist of the production of Yonne, the Côte d'Or, the Côte de Nuit, the Côte de Beaune, and Beaujolais. The following chart lists the major Appellations Contrôlées of Burgundy.

Red Wines of Burgundy

There are two major "types" of Burgundy red wine:

1. The classic long-lived red Burgundies, generally fuller, more velvety, more luscious, and more fruity than the Bordeaux wines, are typically and predominantly made from the *Pinot Noir* grape.
2. The popular short-lived, brilliant, light, fresh, vigorous, fruity Beaujolais is made from the *Gamay* grape.

The classic red wines come principally from the Côte d'Or region, specifically the Côte de Nuit. The Côte de Beaune contributes light wines that mature quickly. The leading wines might be listed as follows, although no official classification is possible:

♦ From *Côte de Nuit.* Romanée-Conti, Chambertin Clos-de-Bèze, Chambertin, Charmes-Chambertin, Clos de Vougeot, Richebourg, la Romanée, Romanée-Saint-Vivant, La Tâche, Musigny, Clos de Tarte, Bonnes Mares.
♦ From *Côte de Beaune.* Château Corton-Grancey, Corton, Corton-Bressandes, Corton Clos du Roi, and Aloxe-Corton.

Beaujolais wines differ substantially from the classic red Burgundies. They are made of Gamay grapes rather than Pinot Noir, and different soil (granitic) and climatic conditions (warmer) have influenced the wine dramatically. The nine best producers of Beaujolais may use the name Burgundy on their labels but often prefer to use the names of the wine vineyard areas (see preceding chart). The appearance of any of these names on a label indicates a potentially excellent wine. The second classification is labeled *Beaujolais-Villages* and must come from designated parishes or villages. *Beaujolais Supérieur*, the third classification, may not be superior at all, and Beaujolais with no qualifier will be a pleasant light wine.

MAJOR APPELLATIONS CONTRÔLÉES OF BURGUNDY

Yonne and Côte d'Or

Bourgogne
Bourgogne Grand Ordinaire
Bourgogne Mousseux

Bourgogne-Aligote
Bourgogne-Passe-Tour-Grains

Yonne only

Chablis Grand Cru
Chablis

Chablis Premier Cru
Petit Chablis

Côte d'Or only

Bonnes-Mares*
Bourgogne Rosé Marsannay
Chambertin*
Chambertin-Clos de Bèze*
Chambolle-Musigny**
Chapelle-Chambertin*
Charmes-Chambertin*
Clos de la Roche*
Clos Saint-Denis*
Clos de Tarte*
Clos de Vougeot*
Côtes de Nuits
Échezeaux*
Fixin**
Gevrey-Chambertin**
Grands Échezeaux*

Griotte-Chambertin*
La Tâche*
Latricières-Chambertin*
Mazis Chambertin*
Mazoyères-Chambertin
Morey-Saintintint-Denis**
Musigny*
Nuits-Saint-Georges**
Richebourg*
Romanée*
Romanée-Conti*
Romanée-Saint-Vivant
Ruchottes-Chambertin
Vosne-Romanée**
Vougeot**

Côte de Beaune

Aloxe-Corton**
Auxey Duresses**
Bâtard Montrachet
Beaune**
Bienvenues-Bâtard-Montrachet
Chassagne-Montrachet
Chevalier-Montrachet
Chorey-les-Beaune
Corton*
Corton-Charlemagne
Côtes de Beaune

Côtes de Beaune-Villages
Criots-Bâtard-Montrachet
Ladoix
Meursault-Blagny
Montrachet
Saint-Aubin
Saint-Romain
Santenay**
Savigny**
Volnay**

Beaujolais

Côtes de Brouilly
Fleurie
Juliénas
Brouilly
Chénas

Chiroubles
Morgon
Moulin-à-Vent
Saint-Amour

Côte Chalonnaise

Givrey	Montagny
Mercurey	Rully

Maconnais

Pouilly-Fuissé	Pouilly-Vinzelles
Pouilly-Loché	

*Leading vineyards.
**Leading production areas.

White Burgundies

There are four distinct families of white Burgundy: *Chablis, Montrachet, Meursault,* and *Pouilly.* Chablis is the most famous, Montrachet considered the best.

Chablis

Chablis, from the northernmost part of Burgundy, is distinguished by a crisp, fresh taste that is favorably likened to a "flinty" taste, or the penetrating taste of metal touched with the tongue on a cold day. In great years, this flintiness is apparent in almost all Chablis, even the lesser growths. There are four qualities of Chablis: (1) *Chablis Grand Cru* (the best), (2) *Chablis Premier Cru* (very excellent), (3) *Chablis* (excellent), and (4) *Petit Chablis* (very good).

While Chablis Grand Cru, Chablis Premier Cru, and Chablis closely resemble each other in quality, Petit Chablis is much lighter and matures more rapidly. By law, all Chablis must be made from the Chardonnay grape. Chablis are not named after producing communes or villages, but the better ones are identified by the names of vineyards. Chablis were officially classified in 1936. The official 1936 Chablis Classifications are listed in the following chart.

Montrachet

The small vineyard of Montrachet in the Côte de Beaune area produces France's most extraordinary white wine—luscious, elegant, full-bodied yet delicate. Montrachet and the wines that have taken its name are much fuller and more flavorful than Chablis. Bâtard-Montrachet and Chevalier-Montrachet vineyards adjoin Montrachet vineyard, and the wines resemble each other. Other wines that merit the name are Bienvenues-Bâtard-Montrachet, Criots-Bâtard-Montrachet, Les Caillerets (formerly Les Demoiselles), Les Combettes, and Les Folatières.

CHABLIS CLASSIFICATION

Grand Cru

Blanchots	Vaudésir
Valmur	Les Preuses
Les Clos	Bougros
Grenouilles	

Premier Cru (in unofficial order of quality)

Montée de Tonnerre	Beugnons
Chapelot	Séché
Mont de Milieu	Châtains
Vaulorent	Côte de Léchet
Vaucoupin	Mélinots
Côte de Fontenay	Les Lys
Fourchaume	Beaugnons
Les Forêts	Troeme
Montmains	Vosgros
Butteaux	Morein
Les Fourneaux	Les Epinottes
Côtes des Prés-Girots	Ronclères
Vaillons	

Meursault

In addition to producing excellent red wines, Meursault produces green-white soft white wines of extremely good quality, although some claim less luscious than the Montrachet: Clos des Perrières, Les Perrières, Les Genevrières, Les Charmes, Les *Santenots*, and others. In some years, however, there are those that claim the Meursault and Montrachet are equals, some claim Meursault surpasses Montrachet!

Excellent white wines are also produced by the vineyard areas of Aloxe-Corton, which are somewhat overshadowed by the red wines of the same region: among the notable exceptions are Corton Charlemagne and Le Corton Blanc.

Pouilly

Maconnais produces a number of notable white wines, but the most distinguished come from the Pouilly (actually, the name of a hamlet), which offers *Pouilly-Fuissé, Pouilly-Vinzelles, Solutré-Pouilly,* and *Pouilly-Loché.* Note that there are also Pouilly wines of the Loire Valley (to be discussed), which are a very different wine. Most of this wine is seen as the golden, heady, rich Pouilly-Fuissé from the communes of Fuissé, Solutré, Chaintré, and Vergisson.

Champagne

Champagne, the world's premier sparkling wine, is unusual in several respects. It is a white wine but is made from black and white grapes, usually in the proportion of 3:1 black (Pinot Noir and Pinot Meunier) to white (Chardonnay) grapes. It is purchased or classified not on the basis of communes, villages, or vineyards, but by brand. Even in excellent years, it is legitimately a blend of several different vineyards' production and several different years.

Champagne is produced by about 200 firms. Of those, only about 30 brands are imported into the United States. The major producing firms own vineyards or contract for the production of grapes in various parts of the Champagne region to produce a wine with qualities that will consistently be identified with the brand name.

Appellation Contrôlée laws, in addition to defining the areas of production of Champagne (of French origin), the grapes used, the quantity of grapes each vineyard can produce, and so on, have regulated its carbonation in the bottle. The law also regulates the production of vintage Champagne. It allows a shipper to blend some wine of other years into a bottle that will be labeled with a vintage year but prevents the production of more "vintage" bottles by blending. Nor can a winemaker sell more than 80 percent of the production as vintage, whether blended or not, or ship it before 3 years on the lees, producing carbonation and full-bodied flavor. In effect, these regulations and the general professional vigilance of an association of producers limit the possibility of a single shipper fraudulently identifying a wine as a vintage. Vintage champagnes sell at considerable premiums.

In most years, Champagnes are all blends of good and ordinary years; they differ first by brand and then by the "degree of sweetness," or how much sugar syrup is added to the dry, carbonated wine before shipping. When the bottle is opened after the second (bottle) fermentation, sugar syrup is added in various quantities to produce *Brut* (very dry), *Extra* Sec (somewhat dry), *Sec* (dry, or somewhat sweet), *Demi-Sec* (sweet) or *Doux* (very sweet).

The preferred taste for degree of sweetness seems to be a national characteristic. Brut is favored by the English; the French prefer Extra Sec or Sec; and Latin Americans favor the most sweet. Presently, even Doux Champagne is only half as sweet as the Champagne sometimes offered 50 years ago.

Most Champagne is white despite being made from black and white grapes. This occurs by preventing fermentation while the grape skins are in contact with the grape juice, as it is the skins that release pigmentation. Pink Champagne can be made by allowing some alcohol from fermentation to dissolve the pigment in the black skins. Alternatively, red wine can be used to dissolve the cane sugar which is added just before shipping, adding the degree of sweetness and rose coloring. Or red wine can be added to obtain the degree of color desired before the dosage is added.

Champagne labeled *Blanc de Blancs* is made from white grapes only; it is usually lighter and drier and is usually Brut. Champagne labeled *Blanc de Noirs* is made from black grapes only. It is full-bodied and hearty compared to a Blanc de Blanc.

Rhône Valley

The Rhône Valley production area is unlike the other wine-producing regions of France. It extends from the Swiss border to the coast near Marseilles. A great deal of wine is produced, including some excellent red wines, some white wines, and some rosés. Part of the production is classified as Appellation Contrôlée, some as V.D.Q.S., but most of the production remains unclassified.

As the price of even the more general Burgundy and Bordeaux Appellation wines increases, wines labeled *Côte du Rhone A.O.C.* appear to be bargains. While they are not Burgundies or Bordeaux, the red Côtes du Rhône are full-bodied, soft when mature, and quite flavorful.

The better wines of the region aspire to more than the simple Côte du Rhone A.O.C. Several red wines are notable: *Côte Rôtie, Châteauneuf-du-Pape, Hermitage,* and *Tavel,* a rosé.

Côte Rôtie

Côte Rôtie produces a fine full-bodied red wine, with deep color and the distinctive taste of *Syrah,* the dominant grape variety of the region. Only a small quantity is produced by all the vineyards in the Côte Rôtie area, and it is not readily available.

Châteauneuf-du-Pape

The Châteauneuf-du-Pape area is the most extensive region of the Rhône Valley. It produces over a million gallons of wine annually. Each producer, every year, sets out to duplicate the distinctive taste of Châteauneuf-du-Pape by combining various grape varieties. The grapes and percentages of each type of grape differ each year depending on rain, the amount of sun that year, and so on. The resulting wine should be consistent from year to year. To qualify as

Châteauneuf-du-Pape, the mixture must be from any or all of 13 different grape varieties designated by law.

The wine is rich, full-bodied, crimson, and relatively high in alcohol (13 percent). Because numerous producers in the area are entitled to the Châteauneuf-du-Pape A.O.C., the name can be used for very different wines, blended of different grapes to the producer's taste. The best wines are produced by single estates: Domaine de la Nerthe, Domaine de Mont-Redon, Cabrières-les-Silex, Château Fortia, Domaine de Nayls, Saint-Patrice, and others.

Hermitage (Ermitage) and Crozes-Hermitage

Hermitage and Crozes-Hermitage are adjoining vineyard areas, producing both red and white wines. Although the wines are similar, the Hermitage area is better located for grape growing, and the wine is a little better. Once these wines were thought to equal or exceed the best Bordeaux or claret, but they never fully recovered from the *Phylloxera* infestation.

This full-bodied, dark, almost purple wine is the most famous of the Côte du Rhône. Unfortunately, production of both the red and the less notable white is limited. Most Hermitage is sold under that name, perhaps with a brand name or shipper's name on the label.

Tavel

Tavel, made from a mixture of grapes dominated by *Grenache,* is perhaps the best and certainly the most famous rosé wine of France. Similar pale, fruity rosé wines are also produced in Lirac and Chusclan.

Tavel wines have a brilliant color and a dryness that is unusual for a rosé. Most Tavel, in the Rhône Valley tradition, is sold by cooperatives, but some is available under vineyard names.

Condrieu

Condrieu is a less well known region of the Rhône Valley with its own A.O.C. Its most famous product is an excellent dry or semidry golden white wine. In some years, it is slightly sparkling. Limited production makes it a rarity outside the Rhône.

Loire Valley

The vineyard area bordering the 600-mile-long Loire River, running basically east-to-west north of Paris, is the longest growing area in France, and the least homogeneous. Vineyards are small, and their production varies in quality from mediocre to great.

There are four main Appellations Contrôlées, each with a clutch of other A.O.C. and some very good V.D.Q.S. associated with it. The name Loire is not an A.O.C., and since only 20 or 25 percent of the wines produced are of any gastronomic interest, this name alone on a bottle means little.

Anjou and Saumur

The west-central part of France produces three types of wine of note:

1. The sweet white dessert wines of Bonnezeaux, Coteaux du Layon, and Coteaux de la Loire, of which that of the Quarts de Chaume is the best, most expensive, and most famous.
2. The Anjou rosés.
3. The white wines of the Saumur.

Some of the wine of the region is sparkling in varying degrees from lightly carbonated (*pétillant*), to wines with as much carbonation as Champagne. A good deal of the wine of this region is exported for the manufacture of sparkling wine in other countries.

Muscadet

The *Appellation Contrôlée Muscadet* refers to a grape, not to a particular growing region. Muscadet's dryness, flintiness, and crispness commend it to admirers of Chablis. The best muscadet wines are produced in the Coteaux de la Loire region of Anjou and labeled as A.O.C.; other producing regions surround the towns of Vallet, Clisson, Vertou, and Saint-Fiacre and bear the Appellation Contrôlée of Muscadet de Sèvre-et-Maine.

Touraine

Touraine has a larger production than Anjou and Saumur. In addition to producing wines that resemble the Anjou rosés and the Saumur whites, it offers the best reds to be found in the Loire: Bourgueil, Chinon, and Saint-Nicolas-de-Bourgueil, all A.O.C.

Vouvray is the most famous Appellation Contrôlée of Touraine. The wine sold under this name can vary from an almost sweet golden white wine to a sparkling wine quite resembling Champagne. While it must be produced in just eight communes to merit the A.O.C., other Appellations Contrôlées of the area usually resemble it in its best-known form, slightly sparkling and crisp: Montlouis, Touraine-Mesland, Touraine Pétillant, and Touraine Mousseux.

Sancerre and Pouilly

Sancerre and Pouilly, vineyard areas on opposite sides of the Loire, are known for their quality flinty whites with great freshness and fruitiness. Among these, Pouilly-Fumé (A.O.C.) is the best known, both because of its quality and because of its long life and ability to travel well.

Pouilly-Fumé is made primarily from the Sauvignon Blanc (or Blanc Fumé) grape, has a bit of a flinty or even, some say, smoky, taste, and is often a pale yellow-green in color, with a touch of green in the meniscus. Contrast this with Pouilly-Fuissé, a white burgundy wine made from the Chardonnay grape, more golden in color and smoother in taste, not as "racy."

Alsace

Unlike the wines of other growing regions in France, wines from Alsace, on the French side of the Rhine, generally take their names from the grape used to produce them. There is only one Appellation Contrôlée, *Vin d'Alsace* or *Alsace*. It always appears larger than the other names on the label.

When a wine from Alsace is labeled with the name of a grape, such as Riesling, the bottle must contain only wine from that grape. Production is subject to the A.O.C. laws and the wine statutes of the area, which are interpreted by a committee of experts who determine the cultivation area, minimum amounts of grape sugar and alcohol, and the harvest date.

A wine containing a mixture of grapes is labeled *Zwicker* (blended) if it contains any grape varieties classified as ordinary, or *Edelzwicker* (finest blend) if it contains only noble varieties. Double names, such as *Pinot-Traminer*, which may appear in the literature, are no longer permitted. Often the name of the grape will be followed by the name of the vineyard where it was produced, or site, or shipper, such as *Riesling Trimbach*.

Mass-Produced Ordinary White Grape Varieties

Chasselas and Sylvaner are widely planted in Alsace. The Chasselas is a light modest wine, white or rosé. It is seldom seen outside of Alsace. Sylvaner can be either ordinary or noble depending on the time of harvest (noble requires a late harvest). Most often, the ordinary Sylvaner is produced in the Bas-Rhin district, the lesser growing region of Alsace.

Knipperle used to be more widely planted but has now been replaced by Chasselas or the Müller-Thurgau, which resembles Chasselas, although it is a cross of Riesling and Sylvaner.

Noble Varieties

The white wines for which Alsace is known are produced from several varieties of grape: the Sylvaner, Traminer, Gewürztraminer (a development of Traminer), Riesling, Pinot Gris or Tokay d'Alsace, Pinot Blanc, and to a much lesser degree Muscat.

The *Sylvaner* is perhaps the least of the noble varieties. Wine from it is pleasantly acid, fruity, and sometimes slightly sparkling. Its strongest advocates describe it as "refreshing." In all fairness, part of its lack of reputation is due to mass production in the Bas-Rhin. The Sylvaner from some areas can be very good.

The *Traminer* or *Gewürztraminer* (spicy Traminer) is the best known of the wines of Alsace. It is highly perfumed, flinty, and dry, but not acid. Traminer is less spicy (less aromatic) than the Gewürztraminer. Once you have tasted Gewürztraminer, you remember its distinctive bouquet.

Riesling is also well known and often more highly praised than Gewürztraminer for its full body, high alcohol content, delicacy, and aroma. In good years, it undoubtedly produces the finest Alsatian wines.

The *Tokay d'Alsace* does not resemble the full-bodied, sweet Hungarian wines of the same name, but rather is a dry or semidry, mellow, slightly pink wine.

Pinot Blanc produced in Alsace is acid and full-bodied. Of all the noble varieties, it is the least distinctive.

Muscat wine from Alsace, unlike Muscat wines from southern Europe or California, is a dry, fruity, perfumed wine.

Alsatian Place Names and Vineyards

Some areas, villages, "sites," or vineyards have gained sufficient reputation to be important on the label. Also, because of the system of classification, the reputation of the shipper is important.

The most important areas of production are Thann, Riquewihr, Ribeauville, Bergheim, Humawihr, Zellenberg, Mittelwihr, Kietzheim, Voegtlinshoffen, Ammerschwihr, Beblenheim, and Turckheim.

Some sites are important enough to stand alone without the name of the communal area: Rangen, Côte d'Olwiller, Kaefferkopf, Kanzlerberg, Schlossberg, Schoenenbourg, Sonnenglanz, and Florimont, among others.

Principal shippers include Leon Beyer, Domaines Dopff, Dopff & Irion, Theodore Faller, Jerome Lorentz, Metz Frères, and F.E. Trimbach.

❖ WINES OF ITALY

In both the production and consumption of wine, Italy exceeds France. France, however, is considered to exceed Italy in quality of wines and in the organization of vineyards and classification. The Italian Wine Laws of 1963 (Presidential Decree No. 93), passed to make production somewhat consistent with Common Market regulation, has the potential of offering the purchaser a classification and labeling system similar to the French Appellation Contrôlée scheme. At the moment, the 1,500 different wines produced by thousands and thousands of small vineyards in a production area of 15,000 square miles are only beginning to be systemized.

Most of the wines, about 96 percent, are consumed in Italy. They are manufactured with the local market in mind and tend to offer vigor, roughness, and acidity rather than mature qualities of delicacy, smoothness, and bouquet. Few Italian wines are aged, but those that are taste quite good.

It is unlikely that consumers in the United States will ever taste the majority of Italian wines. Because American vineyards produce ample quantities of similar ordinary table wines, some made from grapes of Italian origin by vintners of Italian origin, importation is redundant and uneconomical.

Some Italian wines, however, are fine and have the possibility of becoming superb if the efforts of the government and professional associations (*consorzii*) are successful.

The law requires that wines be made from traditional, not hybrid vines. Production standards must conform to approved standards for planting, cultivation, fertilization, maximum allowed yields, maximum allowed residue, minimum alcoholic content, physical and chemical characteristics, as well as origin, if the wine is to be awarded the *Denominazione di Origine Controllata* (D.O.C.) or the more exacting *Denominazione di Origine Controllata e Garantita* (D.O.C.G.). At this time, the D.O.C. or the seal of a consorzio on a label means that the wine in the bottle is worthy of governmental or professional interest and is quite possibly excellent.

There are eleven major production areas:

1. Piedmont and Liguria.
2. Lombardy.
3. Veneto.
4. Trentino-Alto Adige and Friuli-Venezia Giulia.
5. Emilia-Romagna.
6. Tuscany.
7. Umbria and Latium.
8. Marches.
9. Abruzzi.
10. Campania, Lucania, Apulia, and Calabria.
11. Sicily and Sardinia.

Piedmont and Liguria

Piedmont is the most famous Italian wine region; its production exceeds that of the United States and the Russian states combined. The Alps in the north provide natural protection from northerly winds, and vines flourish as far north as Susa and across the plains of Novara and Vercelli to the foothills of Monferrato.

Liguria faces the sea and lies directly to the south of Piedmont. It is better known for its seafood and fish than for its wine. *Barbera*, a fine, rich, full red wine is produced in the provinces of Asti and Cuneo and in the Monferrato area of Piedmont: *Barbera d'Asti, Barbera d'Alba*, and *Barbera del Monferrato*. These wines are made mainly from Barbera grapes. The wine from Asti and Alba is dark ruby in color when young, pomegranate when mature. Barbera del Monferrato is a bright red wine and sometimes slightly sparkling.

Barolo wine, a rich, full-bodied, smooth-when-mature, full-of-tannin-when-young wine is made in the Langhe hills south of Turin. It is produced only from Nebbiolo grapes. It has a rich bouquet of violets and roses, a bright ruby color, full body, and a relatively high alcohol content of 13 percent. Barolo generally matures in 4 years and peaks in about 8; it is very much in demand.

Barbaresco is similar to but not as full-bodied as Barolo; it has slightly less alcohol and matures after 3 or 4 years. *Gattinara* and *Carema*, velvety, robust, pomegranate-red, are also produced from Nebbiolo grapes. The *Nebbiolo d'Alba* dry wines are similar to them. There is also a moderately sweet and a slightly sparkling *Nebbiolo d'Alba*.

Piedmont produces only one still white wine of real note: *Cortese*, a very pleasant straw-colored wine, with moderate alcohol and acid.

Asti Spumante, the Italian "Champagne," is produced in Piedmont from the sweet Muscat grapes grown in the province of Asti. It is much sweeter than Champagne, not from added sugar syrup but from unfermented grape sugar. The production method is also different from that used for true Champagne. While much sweeter than most Champagnes, a well-made Asti can be a wonderful complement to a dessert course or can be served in lieu of dessert.

The white wine of Liguria is probably better known than the red, especially *Cinque Terre,* a dry, slightly fragrant yellow to golden-yellow sweet pleasant wine. *Rossese,* from the grapes of the same name, is a deep pomegranate-red wine, dry, with 13 percent alcohol, and is among the better Ligurian reds.

Lombardy

The wines of Lombardy are mostly made from the Barbera, Nebbiolo, and Croattina grapes. Oltrepo Pavese and the Valtellina are the major production areas.

Oltrepo red wines, made from Barbera and Croattina grapes, include the bright red slightly sweet Barbacarlo, the semisparkling, slightly bitter *Sangue di Giuda,* and the deep red, dry Buttafuoco. Four white varietal wines worthy of mention are also produced: light yellow, aromatic, pleasantly sweet *Bonarda;* light, dry, *Pinot dell'Oltrepo Pavese;* straw-yellow, dry, full-bodied *Riesling dell'Oltrepo Pavese;* and *Cortese,* which is similar to Riesling but softer.

The wines of the Valtellina, produced largely from Nebbiolo grapes, are similar to the wines of Piedmont. Depending on the area where they are produced, they are called *Sassella, Grumello, Inferna,* and *Valgella.* All of them are a little sharp when young, growing softer with age until they are ruby with a dry, full-bodied, well-balanced taste.

The red and rosé wines produced on the shores of Lake Garda are produced from Gropello, Sangiovese, Barbera, and Berzamini grapes. Both the red and the rosé are slightly bitter.

Veneto

Bardolino and *Valpolicella* are the most famous wines of the Veneto region. Bardolino, light red, brilliant, delicate, fresh, and smooth, is produced from several varieties of grapes near Verona. The ruby-red, more robust, slightly bitter Valpolicella is produced north of Verona in the Valpolicella and the Valpantena valleys, from the Corvino, Rondinella, and Molinara grapes.

Recioto Veronese is a special wine made in the same area using only those grapes from the top and sides of the bunch (those grapes receive the most sunlight); they are often late-harvested so the juice has concentrated natural sugars. Alcohol content can reach 14 percent.

The province of Verona produces *Soave,* one of Italy's finest white wines. It is produced from Garganega and Tebbiano grapes in about two dozen villages. Soave has a straw color sometimes slightly green, a moderate alcoholic content, a mellow dryness, and a delightful aroma.

Trentino-Alto Adige and Friuli-Venezia Giulia

A great number of excellent wines are produced in the Trentino-Alto Adige and the Friuli-Venezia Giulia. They are better known in Europe, especially in Switzerland and Germany, than in the United States.

Four wines of the Trentino-Alto Adige merit attention—Santa Maddalena, Lago di Caldaro, Teroldego, and Terlano. *Santa Maddalena* is a superior red wine made from Schiavona, Schiave, and Lagrein grapes. It matures in one year and reaches its peak in two. The color becomes dark orange as it matures.

The soft, ruby, well-balanced *Lago di Caldaro* is another worthy red wine of the region. *Teroldego* has more alcohol than either Lago di Caldaro or Santa Maddalena. It is more robust when young, but gradually becomes smoother, while retaining its characteristic taste of almonds.

Terlano is the region's most famous white wine. It is produced from white Pinot grapes. Its original pale straw color with a greenish tinge changes to gold when it matures.

The wines of Friuli-Venezia Giulia often have foreign varietal names such as Cabernet, Cabernet-Sauvignon, and Tocai. Cabernet resembles its French cousins. White Cabernet-Sauvignon is straw-yellow, high in alcohol, with a slightly bitter, warm taste and a delicate aroma.

Tocai is lemon-yellow, dry, velvety, and mellow, unlike the Hungarian Tokay dessert wines.

Emilia-Romagna

Although Emilia-Romagna is among the largest Italian regions, it does not produce many wines that are famous in the United States. Only the red, slightly sparkling, sweet *Lambrusco* is really well known. The Lambruscos are named after areas of production, such as Lambrusco Salamino, Lambrusco Grasparossa, or Lambrusco di Sorbara. Another red wine, *Sangiovese,* is also worthy of note. Made from the Sangiovese grapes, it is deep ruby with a purple tinge and has a dry, slightly bitter taste.

Only one quality white wine is made in Emilia-Romagna: *Albana di Romagna,* made from the Albana grapes from Bologna to Ravenna. There are dry, semisweet, and slightly sparkling versions. The dry Albana has a well-balanced, slightly acid taste; the others have a sweet, fruity flavor.

Tuscany

Chianti is Tuscany's and perhaps Italy's most famous wine. There are three different types, all produced from Sangiovese, Canaiolo, Malvasia, and Trebbiano grapes. Mass-produced Chianti, for local and Italian consumption, is made to be drunk young. It is low in alcohol, bright ruby-red, dry, rough, and intense. The second type, traditionally exported in the flasks covered with straw, has been aged to acquire more body. It is garnet red and more perfumed. Lastly, there are the more full-bodied, well-aged Chiantis in traditional Bordeaux bottles, higher in alcohol, and a warm, rounded, mellow, exquisite taste.

Although the Chianti wines are produced over a wide region and bear the names of many areas—for example, Chianti Montalbano, Chianti Colli Fiorentini, Chianti Colli Pisane—the Chianti Classico region produces the best wines. Chianti Classico wines are identified by a rooster emblem on the neck of the bottle.

Vernaccia di San Gimignano, light golden-yellow, dry, well balanced, bitter, is Tuscany's most famous white wine.

Umbria and Latium

Two wines of Umbria are worthy of attention: the sweet or dry pale golden, delicate *Orvieto* and the greenish-yellow, dry, slightly astringent *Trebbiano.*

To the south, Latium produces a number of good white wines. The pale yellow *Est! Est!* of Montefiascone is well known for its amusing name and its inherent qualities. The dry version is full-bodied.

The other wines of Latium are known as the Castelli Romani wines, of which each town has its own. The better-known ones are Marino, Colli Albani, Frascati, and Torgiano. *Frascati,* made from

Mavasia and Trebbiano grapes, is most famous; it is straw-yellow, savory, soft, and velvety. *Marino,* on the other hand, made from only Malvasia grapes, is more dry and fruity than Frascati. *Colli Albani* is dry or slightly sweet and has a noticeably fruity taste. *Torgiano Rosso* is ruby-red, dry, well-balanced, and full-bodied.

Three generally good varietal wines are found in Latium: pale straw-yellow, delicate Trebbiano, a dry rosé called Sangiovese, and a deep red, slightly bitter wine called Merlot.

Marches

Verdicchio di Jesi is the best-known wine of the Marches region. It has a delicate, light straw color with greenish glints and a dry, well-balanced taste with a slightly bitter aftertaste. Verdicchio made in the southernmost part of the region can be called Verdicchio Classico. Both are marketed in bottles shaped like ancient amphoras.

Two good red wines produced in the Marches region are ruby-red, soft, full-bodied *Rosso Conero* and the similar *Rosso Piceno.*

Abruzzi

The Abruzzi region produces notable wine, from the Montepulciano grape. *Montepulciano* is ruby-red with a pronounced dry taste and a pleasant aroma. After several years of aging, it acquires more body and mellowness.

Campania, Lucania, Apulia, and Calabria

The wines of the islands of Capri and Ischia are among the better-known wines of Campania. *Capri Bianco* has a pale straw color and a dry, fresh, pleasant taste, with an alcohol content of about 13 percent.

Ischia produces a white and a red, full-bodied dry wine, as well as the ancient Falernum, presently called *Falerno.* The red Falerno is austere, full-bodied, and ruby-red. The white is pale straw-yellow with amber glints.

The mainland produces red and white *Lacrima Christi.* The white is pale straw with an amber tinge and has a well-balanced taste and subtle aroma. The red has deep ruby color, a dry taste, and full bouquet.

ITALIAN WINES

Name	Description
Piedmont	
Asti Spumanti	Brilliant, straw color, delicate bouquet, fresh flavor, sparkling abundant foam
Barbaresco	Brilliant ruby color, delicate flavor, fragrant, dry, more mellow than Barolo
Barbera	Dark ruby, full-bodied, bouquet between cherry and violet
Barolo	Brilliant ruby; after seven years takes on an orange brick color; bouquet of violets and a resinous aftertaste
Bonarda	Dark ruby, sometimes slightly sweet
Brachetto	Bright ruby, pleasant bouquet, sweet and velvety; also sparkling version
Caluso (Passito di Caluso)	Gold or amber; generous perfume, sweet and liqueurlike; rarely dry
Chiaretto di Viverone	Pinkish-white, semidry, nutty
Cortese	Pale straw with green glints; delicate, tart
Fara	Ruby, semidry
Freisa	Deep purple, raspberry bouquet, sharp flavor; also sweet and sparkling
Gattinara	Deep ruby, astringent; mellows with age
Grignolino	Purple, clear, brilliant, nutty; mellows with age
Moscato d'Asti	Straw, slight muscatel perfume, sweet, sparkling
Nebbiolo	Light ruby when young, orange-pink when mature; also sparkling; also rich ruby like Barolo
Rosso Rubino di Viverone	Ruby, very dry
Vermouth di Turino	Both red and white
Liguria	
Cinque Terre	Gold, aromatic, delicate
Coronata	Pale yellow, sweet or dry
Cortese di Liguria	Straw, dry, delicate
Dolceacqua	Light ruby, slightly sweet
Folcevera	Light straw, nutty, sweet; sometimes dry
Portofino	Lemon-yellow, dry
Rossese	Deep pomegranate-red dry wine
Sarticola	Straw, dry, distinctive
Vermentino Ligure	Pale yellow, dry; sometimes slightly sparkling

Name	*Description*
Lombardy	
Barbacarlo	Strong red, delicate, mellow, soft, fruity
Bonarda	Light yellow aromatic wine, somewhat sweet
Buttafuoco	Red, full-bodied, lively
Chiaretto del Garda	Pinkish, smell of almonds, nutty, dry
Cortese	A "softer" Riesling-type wine
Frecciarossa	Amber, semidry, also dry white; also dry rosé; also ruby
Inferno	Deep ruby, distinctive bouquet, soft
Moscato di Castiggio	Pale straw, delicate, sweet; usually sparkling
Rosso Riviera del Garda	Brilliant, light ruby, fruity, nutty, dry
Sangue di Giuda	Dark red, nutty, sweet, sparkling; a little sharp
Sassella	Ruby-red, bouquet of roses, delicate
Tocai del Garda	Brilliant, yellow-green
Villa	Purple-red, very dry
Veneto	
Barbarano	Straw, dry; also ruby, very dry
Bardolino	Light ruby, light body, dry, crisp finish
Bianco di Conegliano	Golden yellow, flinty, dry
Breganze	Straw, nutty, dry; also red, dry, slightly tart
Carbernet di Treviso	Ruby, full-bodied
Cabernet Franc	Ruby, herbal bouquet, flinty
Gambellara	Straw, dry
Moscato di Arqua	Dark yellow, aromatic, sweet
Prosecco	Gold, fruity, tart; also sparkling
Raboso	Ruby, delicate taste of cherries, tart, robust
Recioto Amarone	Ruby, dry, robust
Sóave	Pale amber, velvety, tart
Tocai	Pale straw, delicate, tart
Valpantena	Ruby, distinctive, dry, nutty
Valpolicella	Dark ruby, light bouquet
Verdiso	Straw, tart
Trentino-Alto Adige	
Casteller	Ruby, aromatic, well-balanced
Castelli Mezzocorona	Ruby, dry, full-bodied, well-balanced
Eppaner Justiner-Appiano	Red, dry, slightly bitter, full-bodied

Name	Description
Lago di Caldaro	Light garnet to brick red when aged, full-bodied, mellow, well-balanced
Nosiola	Gold, dry, full aroma
Santa Maddalena	Dark ruby, well-balanced, smooth
Terlano	Brilliant, pale straw, dry; fresh taste
Teroldego	Violet-red, bouquet of almonds and raspberries, matures dry, full-bodied, intense

Friuli-Venezia-Giulia

Name	Description
Bianco del Colli	Straw to gold, dry, slightly sparkling
Bianco del Collio	Straw, dry
Merlot	Deep ruby, dry, smooth
Piccolit	Deep straw, sweet
Pinot	Pale gold, dry, tangy
Pinot grigio	Straw to gray-pink, sharp
Refosco	Dry, ruby, fruity, a little flinty
Sauvignon	Straw, tart
Terrano del Carso	Bright red, nutty, raspberry bouquet
Tocai del Collio	Lemon-yellow, full-bodied, dry
Tocai Friulano	Lemon-yellow to pale green, dry
Verduzzo	Gold, sweet

Emilia-Romagna

Name	Description
Albana	Gold, mellow, sweet; sometimes sparkling
Bianco di Scandiano	Straw, dry; also sweet; also sparkling
Lambrusco di Castelvetro	Dark ruby, violet bouquet, fruity, slightly sparkling
Lambrusco Grasparossa	Very dark ruby, violet bouquet, astringent, slightly sparkling
Lambrusco Salamino	Ruby, mellow, fruity, sparkling
Lambrusco di Sorbara	Ruby, violet bouquet, fruity, sparkling; dry or sweet
Sangiovese	Dark ruby, dry, tart
Trebbiano	Straw, dry; light

Tuscany

Name	Description
Aleatico di Portoferraio	Deep ruby, mild, aromatic, sweet
Ansonica	Gold, dry, nutty
Arbia	Straw, fine bouquet, dry, tart aftertaste
Bianco Vergine dei Colli Aretini	Brilliant, light straw, dry

Name	Description
Brunello di Montalcino	Brilliant, purple-red, violet bouquet, dry
Candia	Straw, sweet, aromatic; also red, full-bodied, sweet
Chianti Classico	Ruby, mellow, light, tart, violet bouquet
Chianti Colli Aretini	Ruby, dry; often sparkling
Chianti Colli Fiorentini	Ruby, violet bouquet
Chianti Colli Senensi	Light ruby, mildly astringent
Chianti Colli Pisane	Ruby, rough
Chianti Montalliano	Dark ruby, violet bouquet
Chianti Rufina	Ruby, nutty, sparkling
Moscatello di Montalcino	Straw, muscatel bouquet
Moscato d'Elba	Brilliant gold, full-bodied
Procanico	Pale, straw, clear, bright, dry, delicate
Ugolino biserno	Pale straw, dry
Vernaccia di San Gimignano	Straw or gold, dry, tart aftertaste
Vino Nobile di Montepulciano	Ruby, dry, flinty, rough
Vino Santo Toscano	Amber or gold, mellow
Umbria	
Greco	Amber, delicate, slightly sweet
Orvieto	Pale gold, dry or semidry, tart aftertaste. Best of the Umbrian wines.
Trebbiano	Light white wine
Latium	
Aleatico di Gradoli	Amber, mellow, sweet, liqueurlike
Cannellino di Frascati	Gold, tart, dry or semidry
Cesanese	Red, dry or semidry
Colli Albani	Straw, dry, slightly tart
Colli Lanuvini	Gold, full-bodied, flinty
Est! Est! Est!	Light yellow, dry or semidry
Falerno	Straw, dry; also red
Grottaferrata	Gold, dry, slightly sharp
Malvasiad Grottaferrata	Dark gold, sweetish
Marino	Gold, slight tart aftertaste; dry or semidry
Montecompatri	Gold, dry
Moscato di Terracina	Straw, sweet, muscatel bouquet
Velletri	Pale straw, dry; also sweet; also red
Zagarolo	Amber, dry, full-bodied, flinty

Name	*Description*
Marches	
Montepulciano Piceno	Brilliant dark ruby, dry
Rosso Montesanto	Ruby, fruity
Rosso Piceno	Pale ruby, full-bodied, tart
Verdicchio di Jesi	Amber, straw, dry or semidry. Best of the Province.
Abruzzi	
Cerasuolo d'Abruzzo	Cherry-red, rough
Montepulciano d'Abruzzo	Purple-red, fruity, nutty, strong
Trebbiano d'Abruzzo	Dark gold, dry
Campania	
Anglianico	Brilliant red-purple, naturally sparkling
Aspirino	Clear straw, tart
Biancolella	Straw, aromatic
Capri	Pale straw, dry, full-bodied; also red
Epomeo	Straw, dry, full-bodied; also red
Falerno	Straw, dry, delicate bouquet; also dry, fruity red
Forastera	Straw, dry, slightly sparkling
Gragnano	Dark purple, nutty, dry
Ischia	Greenish straw; also dry red
Lacrima Christi	Straw, dry, aromatic; also rosé; also red with bouquet of violets and almonds
Sanginella	Straw with gold tint, sweet, slightly sparkling
Vesuvio	Reddish purple, sparkling, full-bodied, slightly astringent
Apulia	
Aleatico di Puglia	Dull red, mellow, sweet aromatic, liqueurlike
Barletta	Dark purple-red, heady
Castel del Monte	Dark ruby, dry, flinty; also dry white; also dry rosé
Malvasia Bianca	Gold, sweet, soft, aromatic
Mistella	Ruby, orange, full
Moscato del Salento	Amber-gold, warm, smooth, mellow
Moscato di Trani	Gold, rose bouquet
Primitivo di Gioia	Vivid red, sparkling, full-bodied; also sweet
Rosato del Salento	Reddish pink, dry
Torre Giulia	Dark yellow, dry
Torre Quarto	Purple-red, dry
Zagarese	Dark ruby, sweet flavor, delicate

Name	Description
Lucania	
Aglianico del Vulture	Purple-red, fruity, naturally sparkling
Malvasia del Vulture	Straw, sparkling
Moscato del Vulture	Straw, sweet, sparkling
Calabria	
Balbino d'Altromontone	Straw, fruity, liqueurlike
Ciro di Calabria	Dark ruby, lively flavor; also sweet
Greco di Gerace	Gold or amber, orange blossom bouquet
Lacrima	Dark ruby, dry
Moscato di Consenza	Amber, sweet
Pellaro	Red to pink, dry
Savuto	Ruby, dry, smooth
Sicily	
Albanello de Siracusa	Gold, dry, warm, full-bodied, mellow, well-aged
Capo Bianco	Pale straw, delicate, dry, smooth; also red Capo
Cerasuolo di Vittoria	Cherry-red when young, almost white after 30 years, full, dry, jasmine bouquet
Eloro	Straw, dry, baked taste; also red
Etna	Greenish white, dry, delicate, well-balanced; also rosé and red
Faro	Bright ruby, dry, well-balanced, delicate bouquet of orange blossoms
Malvasia di Lipari	Gold, smooth, sweet, mellow, soft, well-balanced
Mamertino	Gold, dry or sweet
Marsala	Gold, sweet
Sardinia	
Anghelu Ruju	Ruby, rich, full, sweetish cinnamon bouquet
Monica	Ruby to orange, delicate, smooth, warm, well-balanced
Nascu	Gold, dry, slightly bitter, full-bodied
Oliena	Ruby, dry, resinous, taste of strawberries
Vernaccia	Gold, dry, sweetish
Vermentino	Pale straw, dry, slightly bitter, delicate

The Apulia region produces a great quantity of red and white robust wine, much of which is blended or made into vermouth. Some effort has been made to sophisticate the wines. *Sansevero*, red, rosé, or white, for example, and *Locorotondo* have delicacy and quality.

Sicily and Sardinia

The island of Sicily produces the well-known *Marsala* dessert wine, as well as a straw-colored fresh dry *Etna Bianco* and a ruby-red *Etna Rosso* that develops considerable quality when aged.

Sardinia is famous for its dessert wines but also produces a strong golden-yellow wine with a tart taste called *Vernaccia di Oristano* and a ruby-red *Oliena*, as well as full-bodied white *Malvasia*. The chart on pages 318–323 provides an overview of Italian wines.

❖   WINES OF GERMANY

The 1971 vintage wines, released in 1972, were subject to a new labeling and classification system that has been designed to make identification simpler and to provide more quality control of German wines. Progress has been made in two important areas: (1) guarantee of origin and (2) quality identification and guarantee.

Until the new wine law, the only absolute guarantee of authenticity was an estate-labeled and estate-bottled wine. Geographic descriptions were extremely capricious. Wine might be labeled after any village within a certain area, so that in effect identical wine from the same vines might be called by different names. The present law divides Germany into wine-producing districts, and the districts into smaller areas called *Bereichs*. The Bereich is composed of consolidated small vineyards that alone do not meet the minimum acreage and the individual vineyards that are sufficiently large. The consolidated vineyards are called *Grosslagen* and the individual vineyards are *Einzellagen*. Consolidation has resulted in the elimination of over 25,000 names; this alone simplifies identification.

If a wine is now labeled with a village name, for example, Johannisberg, it must come from that village, not merely from the Johannisberg area. When the wine comes not from Johannisberg but from the general vicinity, it is labeled "Bereich Johannisberg."

When it comes from a Grosslage consisting of the vineyards immediately bordering the village of Johannisberg, for example, the Grosslage of Erntebringer, it is labeled with the name of the village and the name of the Grosslage—in this instance, Johannisberger Erntebringer. When it comes from an individual vineyard within the village of Johannisberg, for example, Holle, it is labeled with the name of the village and the name of the vineyard—in this instance Johannisberger Holle.

The law also deals with quality, which depends on the weather, the location of the vineyards, and the time of harvest. The quality of German wines is first dependent on their sugar content, which in turn depends on the amount of sunlight the grapes receive. The amount of sunlight depends on the weather in a particular growing season. It is said that a hundred days of sunlight make acceptable wine, but a hundred and twenty are necessary for great wines. After the weather, the location of the vineyard is important; position on a hillside determines how many hours of sun will be received. Location within a vineyard is also an important consideration, and the wines of the great estates in Germany are bottled from casks that correspond to different sections of the vineyard and different times of harvest. Unlike the practice of other regions, the grapes with the best qualities are not blended in vats with other grapes of the same vineyard to achieve a high middle standard. Excellent grapes are separated.

Besides weather and location, the time of harvest is a factor. The later a specific bunch of grapes is harvested, the more sugar it will have, both by greater exposure to the sun (hence more ripeness) and by the action of noble rot, which intensifies sugar content.

When sugar content is not sufficient, whether due to poor weather, early harvest, or any other reason, sugar can be added so that the grape juice has sufficient sugar to be made into wine.

The law organizes these variables and documents them on the labels, grouping wines into quality classes that are dependent on chemical, physical, and the organoleptic qualities of the wine, not on geographic location. In other words, the same authentic geographic location, vineyard, consolidated vineyard, or subregion could produce—and would produce—wines of different qualities that would be so identified in the same year. The other qualifier is the cask number so that bottles from casks of different characteristics, which are from the same vineyard, in the same year, and of the same general quality, are identifiable.

There are three quality classes:

1. German table wine.
2. German quality wine of designated regions.
3. German quality wine with special attributes.

German Table Wine (Deutscher Tafelwein)

This is locally grown, inexpensive table wine, or jug wine.

These table wines come from delineated areas and are made from approved grape varieties, mostly Riesling and Sylvaner. The label shows the name of the region—for example, Moselle, Rhine, or Main—or the village but not a vineyard name.

Quality Wine of Designated Regions (Qualitätswein bestimmter Anbaugebiete)

These wines are above average in quality, produced in designated regions from approved and suitable grape varieties. After the wine has passed a governmental analysis and tasting panel, it is given a control number that appears on the label. It must have reached a minimum alcoholic strength through natural sugar and must conform to the typical taste of the region and the grape. Quality wines must originate from certain geographic areas and may be named after a village or vineyard. Wines of this class always bear the words "Quality Wine" and show the control number on the label.

Quality Wines with Special Attributes (Qualitätswein mit Prädikat)

These wines are of the highest classification. In addition to the labeling information for Quality Wines, different attributes are also indicated:

Kabinett (cabinet) means that the wine is made only from fully matured grapes, without added sugar and from the harvest of a very limited district. Kabinett wines pass more stringent standards than those same wines in the Quality Wine class. They are relatively dry in taste.

Spätlese (late harvest) wines come from grapes that are picked after the completion of the normal harvest, having a higher degree of maturity and sugar, which means more flavor, fruitiness, delicacy, and sweetness.

Auslese (selected harvest) wines are made from only the best bunches of grapes, selected and pressed separately. Some of the grapes have the noble rot, which results in a concentration of sugar and flavor. They are sweeter than Spätlese.

Beerenauslese (berry selection) and *Trockenbeerenauslese* (dry berry selection) have the highest degree of sugar concentration. Both are late harvest hand-picked, berry by berry, with only those berries covered with noble rot picked, so as to ensure sweetness. Trockenbeerenauslese wines are made from grapes that have been left on the vine, withering and concentrating the natural juices, long enough to become near-raisins.

Every quality wine must list the bottler (*Abfüller*) and may list the producer. If the wine is bottled by the actual producer who owns the vineyard, then the label says *Erzeugerabfüllung* (bottled by the producer) and *aus eigenem Lesegut* (from their own grapes).

German Wine-Producing Regions

There are only some 250 to 300 square miles of vineyards, compared with the thousands of France and Italy. The yield of quality wines per vineyard acre is the highest in the world, and German white wines, especially fine Rhine and Moselle wines, are unrivaled.

Most vineyards are planted on the steep hills that climb from major river valleys; specifically, the Rhine and its tributaries, the Nahe and the Main, and the Moselle and its tributaries, the Saar and the Ruwer.

The Rhine Region

The Rhine growing area consists of three major regions that are themselves divided into districts, villages, and vineyards. The regions are the Rheingau, Rheinhessen, and the Palatinate.

Rheingau, the best region of the Rhine, produces superior wines of considerable character, which need time to develop in the bottle. The wines of this region are unmistakable for their bouquet, freshness, and fineness. The village areas have achieved independent fame: Eltville, Erbach, Geisenheim, Hallgarten, Hattenheim, Hochheim, Johannisberg, Kiedrech, Oestrich, Rauenthal, Rüdesheim, Winkel, and Kloster Eberbach. Individual vineyards are also famous; among them are Steinberg (Hattenheim), Schloss Johannisberg (Johannisberg), Baiken (Rauenthal), Schloss Vollrads (Winkel), and Doosberg (Oestrich).

Rheinhessen is south of Rheingau and produces fine, fruity, robust wines that are generally softer than Rheingau wines. Major villages include Oppenheim, Bingen, Nackenheim, Nierstein, Worms. Famous individual vineyards include Liebfrauenstift (Worms), Sacträger (Oppenheim), Schlossberg (Bingen), and Rothenberg (Nackenheim).

Palatinate or *Rheinpfalz* is farther south than Rheinhessen, opposite French Alsace. It produces many of the very rich Beerenauslese and Trockenbeerenauslese wines because of its favorable southern location. In general, its wines are aromatic, rich, mellow, and big. The best-known village areas include Deidesheim, Bad Dürkheim, Forst, Königsbach, Ungstein, and Wachenheim. Among the famous vineyards are Freundstück (Forst), Idig (Königsbach), and Frohnhof (Bad Dürkheim).

Two other regions can be included in the Rhine: *Mittel Rhein* (Middle Rhine), which produces full-bodied vigorous wines consumed locally; and the *Baden* and *Württemberg* production area on the river Neckar. Although Baden-Württenberg accounts for a fifth of the growing region, its considerable production is not generally of export quality.

The Rhine Tributaries Other Than the Moselle

The valleys of the Nahe and the Main rivers, tributaries of the Rhine, are also wine-producing regions. The vineyards along the Nahe River produce a number of different wines, but the best known come from the middle Nahe about 15 miles from where it joins the Rhine at Bingen. The best wine producing areas include Schlossböckelheim, Bad Kreuznach, and Niederhausen. The vineyards of Kupfergrube (Schlossböckelheimer), Kahlenberg and Krönenberg (Bad Kreuznach), and Hermannshöhle (Neiderhausen) are well known.

The most famous wines of the vineyards along the Main River are grown near Würzberg in Franconia, about 80 miles from the confluence of the Main with the Rhine. They are known as *Steinwein* (the only wines of Franconia likely to be seen in the United States). Steinwein is bottled in a flat green bottle called a *Bocksbeutel*, rather than in the elongated Rhine or Moselle bottle. The Sylvaner grape in the stony soil of this region produces a very firm wine.

The Moselle Region

The Moselle region, whose wines are often recognized by the slender green wine bottles, is divided into three production areas: *Obermosel*, *Mittelmosel*, and *Untermosel*. Only the last two produce significant export wines.

Mittlemosel and Untermosel wines are light, elegant, delicate, refreshing, lively, and flowery. They do not age as well as the Rhine wines and are best consumed young.

Better-known village vineyard areas include Bernkastel, Brauneberg, Dhron, Erden, Graach, Piesport, Traben-Trarbach, Trittenheim, Ürzig, Wehlen, and Zeltingen. The most famous vineyards are Sonnenuhr (Wehlen), Juffer (Brauneberg), Himmelreich and Josephshöf (Graach), Goldtröpfchen (Piesport), Doktor (Bernkastel), Königsberg (Traben-Trarbach), and Würzgarten (Ürzig).

Tributaries of the Moselle

The Saar and the Ruwer, tributaries of the Moselle, are known for wine. The best production of the Saar River vineyards is from the village vineyard areas of Ayl and Ockfen. The better-known vineyards of the Ayl include Kupp, Herrenberg, and Neuberg; those of Ockfen include Bockstein, Geisberg, Herrenberg, and Neuwies. In the Wiltingen area is Scharzhofberg, the most famous vineyard of the Saar.

Ruwer wines are known for their acidity and fine bouquet and somewhat "stronger" flavor compared to a middle moselle. They keep for two years—unusually long among Moselle wines. The villages of Avelsbach, Eitelsbach, and Mertesdorf are best known. The most famous wine of the Ruwer is the *Maximin-Gurhauser Herrenberg* (Mertesdorf), seconded by the *Karthäuserhofberg* (Eitelsbach) and *Dom Herrenberg* (Avelsbach).

❖ OTHER WINES OF THE WORLD

Spain and Portugal are well-known for their production of specialty wines, such as Sherry and Port, which are unrivaled. Their table wines, both red and white, are also very drinkable.

Other countries that produce wine in large quantities include South Africa, Australia, Chile, Holland, Malta, Crete, Greece, Hungary, Rumania, and Algeria. More and more of these wines are being imported into various countries as consumers' tastes and quests for new wines to try are developed.

Port

The Port of California and New York only vaguely resembles the genuine quality Port of Portugal. True Port is produced in a limited area of the Duoro Valley about 20 miles from Oporto in Portugal.

There are two main varieties of Port: red and white. The white, made from white grapes, is of negligible commercial importance and is merchandised as an aperitif.

Red Port is the traditional Port. There are three main types: Vintage, Tawny, and Ruby, with a number of subdivisions.

Vintage Port

Vintage Port is the wine of a single excellent year, bottled when it is 18 to 24 months old and allowed to mature 8 to 40 years in the bottle. Since this is necessarily an expensive process, there are several variations that accelerate or abbreviate the procedure.

Old Crusted Port is a blend of the wine of several years. It is left longer in the cask, thereby accelerating the maturation process, before being bottled and finished like Vintage Port.

Late Bottled Vintage Port has been kept in the cask for a protracted period, perhaps 10 to 15 years, then bottled, and given additional bottle age.

Tawny Port

Port that has been allowed to completely mature in the cask and is then bottled, usually as a blend, is called Tawny Port or Fine Old Tawny Port. Its age may exceed 20 years, and it is ready to drink when shipped. In fact, further aging may cause it to deteriorate.

Sometimes, lesser wines are sold as Tawny Port when in fact they are relatively young blends of Ruby Ports and White Ports or partially decolored Ruby Ports.

Ruby Port

Ruby Ports are kept in the cask until mature or almost mature, then bottled. They are the cheapest Ports but offer a freshness and vigor that may be more in keeping with modern tastes.

Ports, with the exception of Vintage Ports, are blends of several years, and they are usually purchased by the brand names of various shippers. These blenders attempt to develop an identity for their product. Among the better-known shippers are Cockburn, Smithes, & Cia., Croft & Co., John Harvey & Sons, and Sandeman & Co.

Sherry

Sherry wine is produced in the Jerez de la Frontera region of Andalucía, Spain. Because Sherry is always a blended wine, specific Sherries are identified by brand names or by the names of the shippers. When additional cases of a brand Sherry are needed, an order is sent to a supplier in Spain who prepares a new batch by blending wines of various years, vineyards, and qualities that have been stored in the warehouse (*bodega*).

Sherries are made in various styles that may or may not be identified on the bottles. These styles are listed in the following chart.

STYLES OF SHERRY WINE	
Name	*Description*
Manzanilla	The driest Sherry, often from the Sanlucar de Barremeda region
Fino	Very dry
Amontillado	Less dry than Fino but still dry
Montilla	Resembles Amontillado, but may indicate that the wine is only from the wines of Montilla district
Golden	Medium sweet
Amoroso	Sweet, usually inexpensive sherry
Brown	Very sweet
Oloroso	Full-flavored, nutty; dry or sweet
Cream	Blended from fino and Oloroso
Milk	Blended from fino and Oloroso
Old East India	Dark, brown Oloroso type of sherry

P·A·R·T V

THE STAFF

Basic Personnel Policies

Guest courtesy is of paramount importance. Hiring personnel who are "people" people, who want to be there, who want to be doing what they are doing is part of providing excellent guest service. Having the resources to support the staff is another component of guest service. If the staff is constantly needing more of one supply or another, whether it is coffee cups, spoons, or garbage bags, it makes their job harder, increases the level of staff frustration, and makes each staff member less prone to smiling and providing that "extra mile" of guest courtesy. Providing reasonable rules and regulations is another component of providing excellent guest service. Employees need to know and understand each element of their job description as well as what the "house rules" and regulations are, where employee entrances and exits are located, and so on to feel secure in their job, know what is expected of them, and perform those duties successfully.

The following section describes several items that employers should review and, if relevant to a particular operation, incorporate into their personnel policies if this has not already been done.

❖ BASIC POLICIES AND PROCEDURES FOR ALL PERSONNEL

Reporting to Work—Leaving from Work

Physically Entering and Leaving the Premises

Employees should enter and leave through the designated employee entrance if there is one (usually the case if there are two or more entrances to an establishment).

The employee entrance should be monitored by a security guard during the changing of shifts (when employees are entering and/or leaving). The security guard should check any and all packages, parcels, duffel bags, gym bags, and purses large enough to conceal items not belonging to the employee. This practice usually dissuades employees from "borrowing" supplies.

The front entrance is usually reserved for guests if there are two or more entrances. There should be a specified "delivery entrance" for all deliveries.

Employees should not walk through the dining room, gift shop, or lounge areas while (1) out of uniform or (2) not on duty, unless the employee is there as a guest. Once in uniform, every employee is representing the establishment.

Arriving Early

Employees are expected to allow sufficient time to change, eat the employee meal, if desired, and take care of personal items (e.g., phone calls) before their scheduled reporting time.

Scheduled Reporting Time

At reporting time, employees should be ready for work. The employee meal should have been consumed, any details regarding schedule changes, vacation time, lockers, paychecks, and so on should have been handled, and the employee should be in uniform, ready to start work.

Sometimes employees will arrive before the beginning of their shift so as to gather supplies needed for their shift. Once supplies are gathered, an employee can begin his or her shift productively, getting off to a good start. Such employees include captains, servers, maître d's, bussers, hosts, and hostesses in food and beverage operations, and room attendants in rooms operations. These employees might arrive and be in uniform a half hour before their shift time so they are ready and prepared, with their station or rooms cart filled with supplies, at the start of their shift.

Signing In, Signing Out

Depending on the establishment's rules, employees may be asked to sign in as soon as they arrive on the property or they may be asked to sign in once they are in uniform and ready to begin work.

They may sign in using one of several methods (e.g., time clock and time card, employee card with a magnetic stripe, thumbprint or voice recognition system, or manual time sheet).

Whichever method is used, employees should follow the established procedures. At most establishments, employees must be in uniform when signing in and when signing out.

It is essential to sign in on time. If an employee is late, the shift may be given to another employee who is present. Everyone suffers when an employee is tardy or late—someone has to cover for the person.

Personally Signing In and Signing Out

Each employee is the only person who can sign himself or herself in or out. No one may sign in or check in for anyone else. Breach of this rule usually results in disciplinary action and/or dismissal. Even asking another employee to do something such as, "Sign me in—I'll only be another few minutes—here's my card" or doing such for another employee is usually considered grounds for disciplinary action.

Leaving the Premises during Working Hours

Once the shift has begun, employees should not leave the premises. If there is an emergency, the employee needs to explain the problem to a supervisor so the station can be covered, and request permission to leave the premises.

Lockers

Many establishments have a changing room, some with lockers and/or showers for employees. Some establishments even have an employee lounge.

If the establishment has a changing room or locker room, employees should treat it with care, keep it neat, and not deface any of the property, walls, or fixtures therein. Lockers, if provided, are for employees only. Often, a nominal refundable (e.g., $5.00) deposit is required for the use of a combination lock on the locker, and lockers are assigned.

Using one type of lock on all lockers permits the locks to be opened with a master key. Having a master key permits management to open employee lockers when necessary (e.g., when an employee simply stops showing up for work and presumably has quit). Abandoned lockers can be opened and the contents removed and saved (for return to the former employee). The lockers can then be cleaned and assigned to other employees.

Preventive maintenance is another consideration. Suppose it's the scheduled time for the building to be fumigated, which requires that all lockers must be emptied. Management typically posts a sign asking all employees to empty their lockers by such-and-such a date so the lockers can be fumigated that night. If an employee is on a leave of absence or otherwise unable to attend work, management can open the locker (usually two managers are required to be present to open any locker), put the contents of the locker into a bag, label the bag with the locker number and employee's name, and take the bag to the office until after the fumigation.

If employees bring their own locks, it is usually at their own risk. If any of the preceding scenarios occur, or other occasions arise requiring the opening of the locker, management usually reserves the right to remove a lock.

Packages

Many establishments have created "house passes" for items to be taken off the premises. A house pass would be issued in addition to the register tape when an employee purchases an item at the gift store or at a house sale. If the item is not purchased but a manager or the chef is permitting some goods to leave the premises (e.g., cookies donated for a community fund-raiser), the manager of the department providing the goods (donation) must sign a house pass.

A house pass should include the following information:

1. A description of the goods being released.
2. The quantity of goods being released.
3. The current date.
4. Dollar value of goods being released. This dollar amount should match the register amount if a cash register receipt is also required.
5. Cash register receipt if the goods were purchased from a store outlet, showing the employee discount, if any.
6. Signature of the manager of the department from which the goods were issued.
7. Signature of issuing person, if not the manager of the department from which the goods were issued.

For instance, an employee purchased a birthday cake for his mother's birthday. The assistant pastry chef obtained the "OK" from the executive pastry chef who completed and signed a House Pass and passed it along to the assistant. The employee paid the cashier for the cake and obtained a dated receipt marked, "For birthday cake." The employee then presented the receipt to the assistant pastry chef who marked the receipt "Issued," signed the receipt, and gave the cake and the signed receipt to the employee. The assistant then signed the House Pass to verify the transaction for the house records.

The assistant pastry chef's initials or signature is required to indicate that the correct item was taken, and that the item left the property with the department's knowledge. Security personnel then collect all House Passes as the employees leave the property.

Scheduling Concerns

Posting the Schedule

Everyone needs a schedule posted that shows who is working, or supposed to be working, when and where. Schedules are usually posted weekly. Most establishments attempt to have the next week's schedule posted at least three days prior to the start of the next workweek.

Ideally, schedules should be planned and posted for a month in advance. If weekdays are constant but weekend duty is rotated, only the weekend duty schedule needs to be updated. Both schedules could be posted in the manager's office and on the employee bulletin board, but only the weekend duty schedule would change on a regular basis.

Even though a schedule is posted prior to the week, employees should review the schedule whenever they report for duty to check for last-minute changes. This is true especially if working in a location where banquets and meeting rooms might be reserved or canceled at the last moment.

By reviewing the schedule daily, employees will be alerted to any such changes in the schedule (e.g., working a banquet tomorrow instead of the à la carte dining room; report at 8 A.M. rather than 10 A.M.). Employees who don't check daily might be late for a shift they didn't even know they had. While most establishments try to notify employees personally if the schedule is changed, sometimes it just doesn't happen.

Many establishments will have earlier reporting times for the busiest shifts. For a restaurant, this might occur on busy brunch weekends because there is so much more "prep work" required. It might also hold true for certain days during the week, especially if there is a holiday.

Substitutions

Substituting, or "covering" for another person's shift, is permitted usually with prior written permission of management.

Most establishments ask for at least 72 hours' advance notice. Both employees, the one requesting the substitution and the one who is agreeing to "fill in," should fill out, sign, and have approved a "Schedule Request Form" or similar form. An example of such a form is shown in Figure 16.1.

Absence, Tardiness

Employees who cannot report for work at the assigned time should notify their immediate supervisor. Calling and speaking to a non-manager (e.g., a cook, server, room attendant, maintenance engineer) is not adequate because there is no way to guarantee that the manager will receive the message. If the manager doesn't receive the message it's as if the employee never called.

Employees should call as early as possible so the manager can change the schedule and find someone to replace the absent worker for that shift.

A doctor's note or hospital receipt may be required if an employee is absent due to sickness. Each establishment establishes guidelines so employees know whether they need documentation after one day of sickness or two days or three days, and so on.

Employees who are going to be absent should follow these procedures:

1. Call as early as possible—as soon as a manager is on duty to take the call.
2. Report the absence to your immediate supervisor if possible. If not, be sure to obtain the name of the manager who took the message.
3. Phone calls to report absences should meet the following time frames:
   Luncheon personnel: Call no later than 10:00 A.M.
   Dinner personnel: Call no later than 2:00 P.M.
   Other functions: Call no later than two hours prior to the shift.

If employees do not call to report an absence, commonly referred to as "NC/NS," (no call, no show) there is usually disciplinary action taken. A suspension will often be issued for the first and second occurrence of being late. Dismissal or other disciplinary action is the normal result if an employee is late a third time. The suspensions may be for only a day or two, or as long as a week. The establishment policies and disciplinary procedures are usually posted or available from a manager or human resources person. If the establishment is a union shop, the union steward will have copies of all disciplinary and arbitration procedures.

REQUEST FOR SCHEDULE CHANGE
OR VACATION

NAME: _____

POSITION: _____

TODAY'S DATE: _____

NATURE OF REQUEST: _____

_____

_____

DAY(S)/DATE(S)/INVOLVED, PERSONS COVERING
SHIFTS:

DAY _____ DATE _____ COVERING _____

DAY _____ DATE _____ COVERING _____

DAY _____ DATE _____ COVERING _____

DAY _____ DATE _____ COVERING _____

DAY _____ DATE _____ COVERING _____

If this is a VACATION REQUEST, please indicate the DAY
and DATE you will be returning to work:

  DAY: _____   DATE: _____

Have you arranged for someone to cover your shift(s)? _____

If so, that person needs to sign below to confirm.

Your signature: _____

Replacement 1's signature: _____

Replacement 2's signature: _____

_____

Management Approval: _____

Management Denial: _____

FIGURE 16.1   Sample schedule request form.

PERSONNEL CONDUCT
❖ STANDARDS

Most establishments pride themselves on high standards in the up-keep and decor of the premises, in the quality and freshness of the food, and in their employees' high personal conduct standards. An outline of general policies for employees' personal conduct at any given establishment follows. These are sample paragraphs that might be incorporated into an employee manual or employee policy statement.

SAMPLE EMPLOYEE MANUAL POLICIES AND REGULATIONS

Professional conduct is expected at all times from each employee while he/she is on the establishment premises, whether arriving for a shift, during the shift, or after a shift.

Professional conduct includes each employee's own personal conduct, as well as interactions with other staff members. It involves cooperation and communication with all personnel.

We (management) believe that each employee is honest and dependable. However, if an employee or some employees demonstrate that they cannot or will not follow established rules and policies, such failure to observe those policies will result in disciplinary action leading to dismissal. These policies and procedures apply to everyone who works in our establishment.

Insubordination

To have a smooth-running establishment and ensure the best in guest service whereby guest expectations can regularly be exceeded, it is necessary for all personnel to obey all directions given by a manager or supervisor (e.g., team captain) immediately and without argument.

If there is any reason to question an instruction, the employee should do so at an appropriate time (e.g., at the end of the shift, or if a banquet, at the end of the banquet). Employees should not "show an attitude" or walk off the job. Just do what is being asked, and address the situation later with the manager, the second-line manager, or even human resources or union grievance personnel.

If directives are not followed, it will be construed as insubordination and disciplinary actions may result.

Illegal Drugs

Illegal drugs are not permitted, condoned, nor tolerated on the premises. Any employee possessing, using, distributing, and/or selling illegal drugs while on the premises will be subject to disciplinary action. If the sale or distribution of illegal drugs is involved, federal, state, and/or local authorities will be notified.

Phone Calls

Personal phone calls should be made from designated pay phones (e.g., those in the locker rooms or employee lounge or by the receiving dock or in the kitchen).

Personal phone calls will not be accepted by the reservations office unless there is an emergency. In cases of an emergency, a message will be taken, and the message will be given to the manager on duty. The manager will then relay the message to you.

Smoking

Observe the smoking rules at all times. No smoking is permitted while on the premises. Smoking is permitted outside, on the receiving dock.

Theft

Theft is considered to be the removal of anything from the premises that is not your property. This includes food, clothing, towels, supplies, equipment, menus, money, liquor, and anything else. If it's not yours, don't take it.

If you violate this rule, however small the article is, your action will be interpreted as theft.

Altercations

Do not shout or raise your voice at any time or anywhere in the restaurant or other public areas of the premises. If you are back of the house, your voice should only be raised when there is background noise and only to a level just loud enough to be heard by the person you are trying to speak to. For instance, if you are in the kitchen and the fans are on, speak only loud enough to be heard over the noise of the fans.

Do not engage in physical fights or "pushing matches" or other physical contact with anyone, customer or co-worker, or deliberately cause harm to a guest or co-worker.

Never address anyone, customer or co-worker, in an abusive way.

Chewing Gum

Do not chew gum, or anything else, while on duty.

Gambling

Gambling, of any kind, is prohibited, at all times, in all areas.

Gifts

Gifts may not be given to customers in the establishment without the prior written approval of the manager on duty. "Gift" applies to food, beverages, supplies, and/or any item given to a customer without prior written permission. Any violation of this rule will be considered theft.

Personal Visits

Most establishments prohibit employees from entertaining friends or relatives while on duty or immediately after a shift.

If an employee wishes to dine at the restaurant on his or her day off, as a regular customer, employees are generally welcome to do so after making reservations in advance, letting a manager know of these plans/reservations, and not causing any distractions among fellow workers from their duties.

Employees may not be permitted to patronize the store/establishment where he or she works if there are multiple locations, including at least one other in a nearby geographic vicinity. Often employees will be asked to visit an alternate location and be extended employee benefits there.

Disloyalty

Never speak of the establishment where you work in a disloyal or derogatory way. If you are not proud to be working at the establishment, you may want to consider whether you'd rather be working somewhere else where you might be happier and project a happier image.

Alcoholic Beverages

Alcoholic beverages are not allowed on the premises. They are not permitted to be purchased, consumed, or in your possession while you are on the premises, including the parking lot, while you are on duty, or before or after your shift while you are on the premises.

Nonadherence may result in the establishment losing its liquor license and disciplinary action against the employee.

If the establishment has an "on premises" liquor license, a statement similar to the following might be included in the manual:

> It is prohibited by law for any employee or customer to take any alcoholic beverage off the premises. Our liquor license allows us to provide liquor to guests on the premises, and only on the premises. Such alcoholic beverages must, once ordered, be consumed or disposed of on our premises.

❖ PAYROLL POLICIES

Workweek–Payday

The workweek is generally one week long, perhaps Saturday through Friday; perhaps Monday through Sunday.

Payday is usually one designated day of the week, perhaps Thursday or Friday, usually after 2:00 P.M. Employees are generally paid for the prior week if paychecks are issued weekly.

Paychecks are usually distributed by either the employee's manager or through the accounting office or bookkeeper if there are tips to report.

Some establishments have check-cashing arrangements with a specific bank, perhaps only one location of that bank. If so, it's usually the location nearest the establishment. The establishment may issue check-cashing cards for this purpose. Since this arrangement is considered to be a service, usually only paychecks from that establishment can be cashed, and only with a validated card (other checks will not be cashed by the bank unless you have an account there).

Nontipped employees usually receive their paycheck from their manager. Tipped employees often receive their paychecks from the accounting office or an office where the check can be issued and a *Form 4070: Employee's Report on Tips* can be completed reflecting tips earned for the current week. Each week's reported tips are taxed in the following week's paycheck.

Tip Reporting Procedure

Federal regulations require that service employees report all gratuities as taxable income and that employers withhold taxes on that basis from service personnel and then forward those monies to the appropriate taxing agency.

Establishments must comply with all federal, state, and city regulations, deducting the appropriate taxes from employees, then paying the various taxing authorities on the employees' behalf.

Paycheck Decoding

Many establishments use codes on employees' paycheck stubs. The codes are usually initials on the paycheck, each designating a debit or credit to wages earned. Some common codes and possible code letters, indicating what those codes might represent, are shown in the following list.

REG   Regular hourly pay rate $\times$ number of hours.

T   Tips.

M   Meals ($1.50 or so per day from gross pay, if meals are deducted).

A   Uniform allowance.
$X.00 if 1–20 (or whatever the guidelines are) hours are worked
$Y.00 if 20–30 (or whatever) hours are worked
$Z.00 if more than X hours worked

B   Banquet hours worked.

Z   Banquet tips (a deduction if the tips were paid out already in cash; an addition if they are added in here and not paid out immediately after the event).

U   Union dues, if a union shop.

X $X.XX for working a double shift, or more than 8 hours on one day.

F   Lock deposit for locker (usually a one-time charge), or locker charge.

H   Holiday pay ("double pay"). Same number of hours $\times$ 2 for that holiday's hours.

❖ EMPLOYEE BENEFITS

Employee Meals

It is customary to provide a meal, usually with hot entrées, various salads, vegetables, and beverages for employees during their working shift. There may be a nominal charge for the meal, covering the cost of the food.

Meals are usually offered once per shift to ensure that every employee, regardless of shift, will have the opportunity to have an employee meal during that shift.

Meals should be eaten when provided, and at no other times. Snacking while on duty is usually not condoned and, for professional reasons, should not be done. The consumption of the establishment's food, other than that provided for employee meals, is not allowed and is often considered theft.

Employee meals should be consumed only in designated areas (e.g., the employee cafeteria, if there is one). Each person should be responsible for returning china, silver, and napery to the warewashing area after eating.

Beverages

Coffee, water, sodas, and juices are often provided for employees. The supplied beverages, and only those beverages, are to be consumed by employees. Consumption of beverages other than those provided by the employer may be considered theft. Often specialty drinks (e.g., milkshakes), if not provided, may be purchased.

Meals as a Guest

Employees are usually permitted to eat and/or drink as a customer in the restaurant within certain limitations. Dining may be permitted if at least three (or four or whatever number management decides on) hours have elapsed between being there as a customer and being there as an employee. Or dining as a customer may be permitted only on an employee's day(s) off or, with prior permission of a manager, if a special occasion arises (mother's birthday or unexpected visit from dad). Usually, however, some provisions have been made so employees can enjoy their own establishment with their friends and family.

Employee Discounts—Gift Shop

If there is a gift shop, employees are often allowed a discount on merchandise purchased there. A typical discount might be 20 percent off retail price. Sometimes, a greater discount is offered on cash purchases; a slightly less discount may be allowed if the purchase is charged on a credit card. Employees may have certain posted "shopping" hours for purchases, often the hour prior to opening to the public.

❖ # OTHER POLICIES AND PROCEDURES

Accidents/Emergencies

In case of an accident or emergency:

1. Inform a manager.
2. Locate someone with CPR and/or first aid and/or EMT training willing to assist.
3. Ask the injured person if he or she knows where and extent of injuries, if any.
4. If none are stated (e.g., a slip on a hardwood floor with no apparent injury and no pain), assist the person up.
5. If the person is injured or *might be injured*, and movement may make the injury worse, wait for medically trained personnel. (The manager who was notified should have called "911," or the police if "911" is not a working police number in that locale, as soon as it was determined that medical aid would be required.)
6. If the manager is not there yet, obtain facts pertaining to the accident:
   - Who: Names and addresses.
   - What happened.
   - When.
   - Where.
   - Why.
   - Witnesses, if any, including yourself, any staff members, and any other guests.
   - Extent of injury, if any, as judged by any medically trained personnel and/or witness(es).
   - Any damage to goods such as a smashed briefcase if a guest fell on briefcase (in case establishment chooses to pay for its replacement); or a ripped dress if a guest fell and ripped her dress on a chair edge; or stains if the accident involved a collision between a server and a guest, during which food or beverages fell from a tray onto the guest.
7. Fill out an Accident Report Form with the ascertained information.
   There are generally two report forms:
   - Public Accident Report Form, for accidents involving guests.
   - Employee Accident Report Form, for accidents involving employees.
8. Ask the persons involved (the guest and/or employees as well as witnesses if any) to sign and date the Accident Report Form.

If it's an emergency (e.g., fire, bomb threat, armed person on the premises, major water leak), most establishments have policies and procedures in place, and provide their employees with the following instructions:

1. Don't panic. Remain calm and notify a manager immediately if possible. Let the manager handle the situation.
2. Do not alarm guests, if possible to avoid doing so.
3. If directed to do so by management, evacuate guests in a calm, orderly manner.
4. Do not play hero. Unless trained, don't touch a suspected bomb, try to talk to or disarm an armed person, or do anything to endanger anyone's life, including your own.

Unless it's an immediate "life-and-death" situation, it's usually best to call the local police precinct telephone number, rather than "911."

Dialing "911" is a matter of public record. Many establishments would rather not have any disturbance be quite so visible. Calling the local number brings a fast response without the notoriety. Post the number to be called conspicuously and in many places (e.g., front desk, by the cashier station).

Generally, it is advisable not to call a doctor if a person is in need of medical attention, but rather to call the police and ask for "police and medical personnel," including an ambulance, to be sent to the establishment as soon as possible. If the establishment calls on a doctor's services, the establishment may be liable for the bill, not the patient. To ensure the establishment is not liable for the medical charges, call the police unless you've been instructed otherwise.

If there is a fire, call the local fire department.

PERSONAL GROOMING AND HYGIENE STANDARDS

Staff members who adhere to the following guidelines will convey a professional image.

Hair

Hair should be clean, well-shaped, well-trimmed, and groomed. There may be length restrictions. Perhaps, for front-of-house food servers, "If it touches the top of your collar, it is too long and should either be cut or pulled back and restrained."

Often for other front-of-house personnel, hair must be well-groomed and if touching or longer than shoulder-length, pulled back or restrained so as not to interfere with duties. At no time should hair be falling into the employee's face (e.g., when serving guests their food).

For kitchen back-of-house personnel, hair must usually be restrained and a cap or hat worn to prevent hair from falling into food preparations and/or getting caught in machinery.

The kitchen back-of-house requirements, sans cap, usually apply for other back-of-house personnel who often work with equipment and machinery.

Facial Hair—Sideburns/Mustache/Beards

Some establishments want a "no facial hair" look for their front-of-house employees, others ask for "neatly trimmed" facial hair, and others require facial and/or long hair (e.g., tourist attractions featuring medieval jousting matches, etc.).

For establishments desiring "neatly trimmed facial hair," the employee manual might state:

> Men should be clean-shaven or with tidy, well-kept facial hair (e.g., sideburns, mustache, beard). The facial hair must be kept clean and well-trimmed. Sideburns, to look tidy, should extend to, and not below, the middle of the ear. A mustache should not extend beyond the "smile lines" for a conservative appearance.

Depending on the establishment and the look desired, both in decor and the costume/uniform of the staff, an establishment might encourage specific facial hair or specific styling (e.g., a "Wild West" bar and tavern might encourage a bartender to create a handlebar mustache and might outfit him in a red-striped shirt with an armband).

Jewelry

Certain types of small, minimal jewelry are often permitted to be worn while on duty. However, since jewelry might become caught in doors, machinery, and so on, it is advised that no jewelry, or minimal jewelry, be worn. The following are fairly standard guidelines.

Earrings

Depending on the theme of the establishment, there may or may not be any guideline or policy regarding earrings (e.g., at a trendy beach resort or flashy nightclub, the wilder the better).

At a conservative restaurant or casino where strict uniform policies are followed, required or conservative earrings may even be issued. If employees are allowed to choose and wear the earrings, they often must be small, discreet, button-type earrings—not dangle earrings, not bright or shiny earrings, and not earrings that display highly visible designs and/or color. Where earrings are allowed to be worn, generally one pair per person is the limit, with one earring per ear.

Wristwatch

A wristwatch is generally permitted—one per person in a conservative establishment. It should be conservative and used to tell time, not as an attraction-getter or conversation piece. In a "funky" establishment, however, employees might be encouraged to wear several splashy watches on one wrist—that's part of the theme. The same establishment might encourage wild earrings or suspenders filled with buttons. Whatever the theme, the dress and accessories should be part of, and not detract from, that theme.

Rings

Wedding rings and/or bands and class rings are generally acceptable as are other small rings with no protrusions that could get caught in clothing, equipment, doorways, and so on.

Hands

Well-manicured hands create a pleasant, professional image. It's important to have clean fingernails, not too long, not too short, always meticulously clean. Fingernails should be short and immaculately clean because whether the employee is a room attendant, receptionist, front desk clerk, food server, or busser or coat-check person, the person's hands are a focal point to the customer. There is nothing worse than being handed a room key or served food by an employee with dirty hands or fingernails.

Nail polish is generally acceptable if all the fingernails are painted one color and the color is clear; neutral, such as a light pink or flesh tone; or a red tone. Whether or not employees in food service should wear nail polish is a constant debate. No one wants a speck of nail polish in their food, yet servers often insist that nail polish helps to harden their nails, allowing them to keep them "professional looking." If servers use nail polish, it should be clear or of a soft or neutral shade. Back-of-the-house food service employees should not wear nail polish at all since they are in daily contact with food and food products.

Whatever the establishment's policy on nail polish, it is imperative that service personnel do not touch hair, nose, or face while in public view—where there is no opportunity to wash hands immediately afterward. Employees should concentrate on what their hands are touching—it should be the menus, the plates, the glasses, a guest's coat, the telephone, the chair to seat someone. Most of the time, a hand touching the face is an unconscious movement (e.g., brushing hair out of the face, which is why it should be restrained; rubbing an itchy nose), but these movements transmit germs and are unsanitary.

Hygiene

Hygiene for most employees is automatic. All employees should follow a daily routine of good hygiene:

♦ Daily bathing.
♦ The washing of one's hair.
♦ Use of underarm deodorant or antiperspirant.
♦ Use of mouthwash.
♦ Daily brushing and flossing of teeth.

Clean undergarments, every day, are essential.

General Appearance

Each employee's general appearance should be professional and in keeping with the guidelines and theme of the establishment. Whatever the theme, employees should always be neat and clean, well-groomed, and wearing a smile.

UNIFORMED SERVICE PERSONNEL

Each establishment, before opening its doors to the public, decides on a theme, with a menu to complement that theme. The decor and costumes, or uniforms, then follow. The prescribed uniforms for staff are usually detailed when each staff member is hired as well as any special stores where employees might be offered "employee discounts" on uniform pieces that they must purchase such as tuxedo pants or jackets.

A conservative, somewhat formal restaurant might prescribe uniforms for the following positions:

1. Maître d', captains, host, coat-check personnel.
2. Servers.
3. Bussers.

The following verbiage from a sample employee manual illustrates the type and level of detail necessary when defining uniform policies so there is no misunderstanding with any employee (see pp. 354–356). Colors must be specified, type of fabric, length of slacks, fit of slacks, and so on. Otherwise, employees will often take creative license and, rather than report to work in black tuxedo slacks, arrive in black stretch slacks with glitter highlights or variations thereof.

Grooming/Appearance Summary

It is important that each employee meet the standards of the establishment because the way employees appear is a direct reflection on the restaurant as a whole. If the table settings are sloppy, the room is sloppy, and the service staff looks sloppy, you've already given the customers the image of a sloppy meal, before they've even seen the food, let alone tasted it! You've let down the customer: Instead of exceeding the customers' expectations—the key to return customers—you've provoked disillusion.

First impressions are lasting, and customers usually form that first impression from a member of the service staff. They are the sales force of the establishment. Each service staff member should convey a message of caring:

◆ About the food—its quality and its preparation.
◆ About the ambience of the facility, the entrance, the rest rooms, the dining rooms, the table settings, each room's appearance.
◆ About his or her own personal appearance.
◆ About the establishment and his or her part in its success.
◆ Most of all, about the establishment's customers.

And remember, establishment rules shouldn't be arbitrary, but have a reason behind them. For instance, at Ronnie's, there is a

SAMPLE UNIFORM VERBIAGE FOR
AN EMPLOYEE MANUAL

The prescribed uniform must be worn whenever working. Usually employees won't be permitted to work if not in proper uniform including the following of good hygiene and grooming procedures. Each uniform must be kept meticulously neat and clean at all times.

Shoes

Black leather, preferably with thick soles to cushion your feet; polished. The shoe should be built well with a solid instep and support for the foot.

Decorative heels are out of the question; as are open-toed or open-heeled shoes or sneaker-style shoes [sneakers might be acceptable in a casual establishment].

Socks/Stockings

Black socks and/or stockings. Socks should have some natural fiber content, either cotton or wool, so they will absorb moisture while you're working, keeping you more comfortable than all-nylon or all-polyester socks.

If you want to wear white athletic socks, wear those, then pull black dress or black athletic socks over the white socks so only the black will show.

Stockings worn with tuxedo slacks should be black or off-black, without holes, snags, and/or runs. Stockings worn with skirts should be taupe, tan, or nude in color, and without holes, runs, snags, or design.

Pants

Pants should be black tuxedo pants, in either wool or a wool/polyester blend. The pants must have the tuxedo stripe running down the length of each pant leg on the outside.

The pants should be conservative in style, hanging straight down and being roomy but not baggy nor tight.

Circumference of pant leg at the ankle (bottom of the pant leg) should be 18 inches (not "wide" or "bell-bottom" pants; not "peg-leg" or tight).

Suspenders

Optional. If worn, a belt should be omitted.

Cummerbunds

Not required. May be issued for certain banquets and special events. Colors will vary depending on the theme and color selection chosen by the client for each event.

Shirt

Servers and Bussers. Dress shirts of 100 percent cotton (or 60% cotton, 40% polyester) with long (or short) sleeves. Each shirt should be pressed and spotless. Yellowed shirts or dirty collars, stained or spotted or unpressed shirts, or ill-fitting shirts are not acceptable. "Button-down" or "oxford" shirts are not acceptable; nor are tuxedo shirts.

Cotton feels better than polyester (it absorbs moisture), looks better, and cleans whiter than polyester. Also, stain removal is easier, and bleaching won't yellow the fabric as fast as it will affect polyester blends.

Maître d', Captains, Hosts, Hostesses. Tuxedo shirt with the previously stated parameters—pressed, neat, clean, without holes or stains, not yellowed. Tuxedo buttons of either white or black. Shirt should be either 100 percent cotton or a 60/40 cotton/poly blend.

Bow Tie

A black, dress, tuxedo bow tie of a size appropriate for your height and stature. Velvet, black-on-black designs, polka dots, and so on are not acceptable except as designated by management (e.g., New Year's Eve or other special holiday occasion). A plain black nonglossy cloth bow tie is the norm.

Aprons

Part of the bussers' uniform. A white, spotless apron is provided by the establishment. It is each busser's responsibility to keep the apron clean during the shift or to obtain a clean one if the issued one becomes soiled.

Jacket

Jackets are issued on employment. A deposit may be required for each jacket. Jackets should be professionally laundered so they remain crisp and spotless.

A monetary allowance is included in paychecks to facilitate professional laundering. If the jackets are home laundered, they must be starched after laundering to maintain the crisp look that is part of our theme.

If your jacket is stained, dirty, yellowed, ripped or torn, you may be sent home and your shift given to another employee in proper uniform. If normal wear and tear is affecting your jacket, have it laundered and exchange it for a new one.

Servers. White, three-button jackets. Provided by the establishment. Available from the accounting office or your manager if you provide the size you need.

A $20.00 deposit is required for each jacket. This deposit, plus interest, is returned to you when the jacket is returned. Should the jacket become stained, and laundering will not remove those stains, the laundered jacket may be exchanged for a new one. You should have at least three jackets so you can rotate them, always keeping a clean, laundered, pressed one available.

Bussers. Black jackets. Provided by the establishment. Available from the accounting office or your manager if you provide the size you need.

A $20.00 deposit is required for each jacket. This deposit, plus interest, is returned to you when the jacket is returned.

Captain, Maître d', Hosts, Coat-Check Personnel. Tuxedo jacket—spotless and black. Professionally cleaned. Each employee provides his or her own tuxedo(s). Ensure that they are tailored to fit. Stain-finish lapels are preferred; avoid velvet or other nontraditional styles.

Pant legs should not touch the floor when you walk; they should be professionally altered and/or hemmed (don't staple or pin the tuxedo together).

Cocktail Servers. A pocketed apron, provided by the establishment is worn with a white dress shirt or blouse and black skirt or tuxedo pants. A black vest, provided by the establishment, is worn over the shirt or blouse.

All items of the uniform should be clean, spotless, pressed, and freshly laundered. Once you have been issued a uniform, you are responsible for keeping it crisp and clean.

---

one-butter-pat-per-person rule. Many customers and even some staff members believe Larry Leckart, the owner, is being stingy or, to give him "the benefit of doubt," cost conscious.

Not so. Larry is, by enforcing this rule, ensuring that his service staff will repeatedly visit each table to see what the guests need, be it more butter pats or coffee or water.

Think about your establishment's rules. Are they customer-service oriented? Will they bring guests back? Do they go "beyond the basics"?

ADA Signage

Information about the Americans with Disabilities Act

The Americans with Disabilities Act (ADA) extends protection of their rights to the disabled under the Civil Rights Act of 1964. The entire text of the ADA can be found in the Federal Registers of July 26 and September 6, 1991. The ADA is a complex set of legislation dealing with many areas of compliance requirements. Although this is Federal legislation under the jurisdiction of the U.S. Attorney General, it is our understanding that each state will create its own adaptation of the law which will then be applied by the local governments.

CAS has prepared this brochure to specifically address the sections of the Act which refer to signage. These requirements are for new construction and alterations (where the compliance is readily achievable) in places of public accommodation and commercial facilities.

Please contact your CAS representative for information on the various ADA Sign Systems CAS has available for use in your building(s). We can also give you additional information on current local codes and all updates to the ADA. As always, we look forward to assisting you with all of your signage requirements: Directories, Identification signs, Exit Plans, Tenant signs, etc.

CAS
THE SIGN SYSTEMS COMPANY

Corporate Headquarters:

10909 Tuxford Street   Telephone: 818/768-7814
Sun Valley, California 91352   Facsimile : 818 504 2944

Reprinted with permission of CAS The Sign Systems Company, Sun Valley, CA.

Directional and Informational Sign Requirements

Typical Directional

Typical Informational

The following requirements are for directional & informational signs for rooms and spaces:

1. Good contrast, light on dark or the reverse, between message and background is required. Please refer to the **CAS** Color brochure for colors.

2. Matte, eggshell or semi-gloss finish. Matte finished brushed metal finishes can be used.

3. Width to height ratio for character is to be in the range from 1:1 to 3:5. (No condensed or extended letterstyles to be used).

4. Width to height ratio for letter stroke is from 1:5 to 1:10. (No very thin or thick stroked letterstyles to be used).

5. Upper and lower case letters are permitted.

6. Character height is to be as shown on the Visibility Chart, use capital height for measuring. For more information, please refer to page 4 in the **CAS** Nuts & Bolts brochure.

7. For overhead signs, the minimum cap height is to be 3". (No braille or raised letters required).

1. & 2.

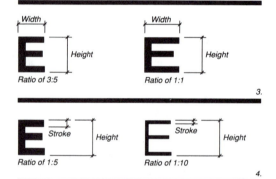

3.

4.

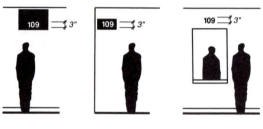

7.

UPPER CASE (ALL CAPS) & lower case

5.

Visibility Chart

*This partial chart is based on **CAS** research under optimum conditions using Helvetica Medium. For a display of actual letter heights up to 6" please see pages 5 and 6 in the **CAS** Nuts & Bolts brochure.*

Letter size in inches:

1/2	1	2	3	4	5	6	10	12
14	28	53	68	82	112	179	450	525

Readable distance in feet

6.

THE SIGN SYSTEMS COMPANY

Corporate Headquarters:

10909 Tuxford Street
Sun Valley, California 91352

Telephone: 818/768-7814
Facsimile : 818 504 2944

Permanent Room & Space Identification Sign Requirements

— Permanent Room Sign

— Non-permanent Temporary Tenant Sign

Typical Permanent Room sign with temporary Tenant sign below

1. & 2.

— Letters are a minimum of 1/32" (.031")

— Braille is .025"

3.a.1 & 3.b

Sans serif letterstyles:

ERAS BOOK *(64)*
HANDEL GOTHIC *(57)*
HELVETICA MEDIUM *(50)*

Simple serif letterstyles:

BERLING *(69)*
OPTIMA SEMI BOLD *(58)*
SERIF GOTHIC BOLD *(66)*

Please see pages 1 & 2 in the CAS Nuts & Bolts brochure for entire alphabets (CAS Code No. in parenthesis) & other letterstyles.

3.a.2 & 3.a.3

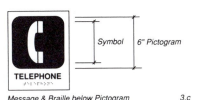

109 — Minimum: 5/8"

109 — Maximum: 2"

3.a.4

Symbol   6" Pictogram

TELEPHONE

Message & Braille below Pictogram   *3.c*

1. Good contrast, light on dark or the reverse, between message and background is required. Please refer to the **CAS** Color brochure for colors.

2. Matte, eggshell or semi-gloss finish. Matte finished brushed metal can be used.

3. Message:
   a. Wording
      1. Letters are to be raised a minimum of 1/32".
      2. Letterstyles are to be sans serif or simple serif .
      3. Letters are to be in all capitals (UPPER CASE).
      4. Cap height is to be 5/8" to 2".
      5. For overhead signs the minimum cap height is to be 3".   (No braille or raised letters required.)
   b. Braille - Grade 2 (189 contractions & short form words.)
   c. Pictograms - 6" minimum border printed with smaller symbol inside with verbal description & braille beneath.

4. Location:
   a. Signs are to be placed on the wall adjacent to the latch side of the door.
   b. The dimension from the floor to the centerline of sign is to be 5'- 0" (60").
   c. For double-doors or if no wall space exists, sign to be placed on the nearest adjacent wall.
   d. Space from door swing or any obstacles is to be a a minimum of 3".
   e. For overhead signs, the clearance is to be 6'-8" (80").

Note: There are additional requirements for Transportation Facilities, for displaying the symbol of access for hearing loss in assembly areas, for areas of rescue assistance instructions and directions, as well as for text telephone (TDD) identification and directions. Please refer to the appropriate Federal Register for information.

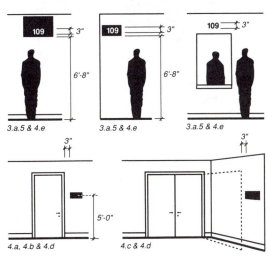

3.a.5 & 4.e   *3.a.5 & 4.e*   *3.a.5 & 4.e*

4.a, 4.b & 4.d   *4.c & 4.d*

CAS
THE SIGN SYSTEMS COMPANY

Corporate Headquarters:

10909 Tuxford Street
Sun Valley, California 91352

Telephone: 818/768-7814
Facsimile : 818 504 2944

Exterior Signs

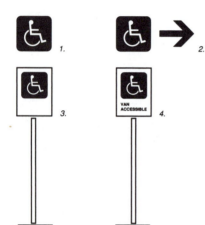

All exterior signs must meet the same requirements as the interior signs, depending on the category to which it belongs - Permanent Room & Space signs or Directional & Information signs.

In addition, the following signs are required for the exterior:
1. Identificaton of accessible entrances.

2. Directions to the nearest accessible entrance.

3. Identification of handicap accessible parking spaces installed so that a parked vehicle will not block the sign. (In California, the sign is to be 70 square inches and located with a clearance of 80".)

4. Identification of handicap Van accessible parking spaces installed so that a parked vehicle will not block the sign. (In California the sign is to be 70 square inches and located with a clearance of 80".)

Signs that do not need to comply with the ADA

Building Directory

Menu			
Wxvkininkl	1.25	Sadhgjk	.65
Tuudsh hsiuhdiuh	2.45	Ljbbbjbj	.75
Fuyg stj ygduy	4.55	Detr uygg	.80
Egg dgg sskljdj	2.50	NJhu	.35
Ruywfuyduvb	3.25	loiuy jjij	.90
Muhsghd sijjid	6.70	Ruygd jkjj	.25

Menu

The information shown on the previous pages also apply to all exterior signs.

The following signs are required for the exterior:
1. Identificaton of accessible entrances.

2. Directions to the nearest accessible entrance.

3. Identification of handicap accessible parking spaces. Install so that parked vehicle will not block the sign. (In California the sign is to be 70 square inches and located with a clearance of 80".)

4. Identification of handicap van accessible parking spaces. Install so that parked vehicle will not block the sign.

Instructions and regulations

The Sign Systems Company

Tenant sign is considered a temporary sign

Russ T. Hinge, Vice President

Name sign is considered a temporary sign

Out of Order

Temporary sign

CAS

THE SIGN SYSTEMS COMPANY

10909 Tuxford Street
Sun Valley, California 91352
Telephone: 818/768-7814
Facsimile : 818 504 2944

Please contact CAS for additional information on our products.

Current and future Brochures:
#1	Neg-Lume Directories	#7	Matrix
#2	Non-illuminated Directories	#8	Plaque & Frame Systems
#3	Nuts and Bolts	#9	Vinyl Graphics
#4	Colors	#10	3D Letters
#5	BullNose	#11	Exterior System
#6	Bevel Edge	#12	Price List

Ullage Charts

Glossary of Abbreviations

Ullages (Level of Wine)

For Bordeaux, Port and other wines in bottles with defined shoulders the ullage/level is shown, if relevant, by its relevant position in the bottle. Our interpretations are as follows:

u. – ullage/ullages

n. – within neck; the normal level of young wines

bn. – bottom neck; completely acceptable for any age of wine

vts. – very top shoulder; completely acceptable for any age of wine

ts. – top shoulder; usual level for wines over 15 years old

hs. – high shoulder; typical reduction through the cork, usually no problem

ms. – mid shoulder; usually some deterioration of the cork and therefore some variation

ls. – low shoulder; more variable and therefore carries a lower estimate

bs. – bottom shoulder; very variable, usually a rare or interesting wine with a low estimate

Example: (u. 3hs) means 3 bottles ullaged to high shoulder.

For Burgundy, German and other wines in bottles with sloping necks the ullage is shown in centimetres, measured from the base of the cork.

Example: (u. 2x5cm) means 2 bottles ullaged 5 centimetres.

Cautionary Notes

1. Whilst we do all that is possible to indicate accurately the levels of older wines, such levels may change between cataloguing and sales. This may be caused by the ageing of the cork or by a change in the temperature of the storage conditions or the shipment of the wine.

2. There is a risk of cork failure in old wines which must be taken into account by the potential buyer.

3. Sotheby's do not, as a matter of policy, inspect original cases.

4. Sotheby's will not entertain any price negotiation or credit after the delivery is made and returns will not be accepted.

BOTTLE SIZES – Quantity of litres per bottle size

			BORDEAUX & BURGUNDY	CHAMPAGNE	PORT
hf.bt.	–	half bottle	0.375	0.375	0.375
imp.pt.	–	imperial pint	0.568	0.568	0.568
hf.ltr.	–	half litre	0.5	0.5	0.5
bt(s).	–	bottle(s)	0.75	0.75	0.75
ltr.	–	litre	1	1	1
mag.	–	magnum	1.5	1.5	1.5
m-j.	–	marie-jeanne	2.5	—	—
d.mag.	–	double magnum	3	—	3
jero.	–	jeroboam	5	3	—
reho.	–	rehoboam	—	4.5	—
imp.	–	imperial	6	—	—
meth.	–	methuselah	—	6	—
salm.	–	salmanazar	—	9	—
balth.	–	balthazar	—	12	—
nebu.	. –	nebuchadnezar	—	15	—

The above table is a guide to the litres of wine per bottle size. Where relevant the figures have been taken from the EEC prescribed litreage for light still wine, sparkling wine and liqueur wine. The different categories have separate implementation dates which, when combined with past variances in bottling quantities of some bottle sizes, means that this should be treated purely as a guide. Should you require the litreage capacity of the wine in any lot, please contact the wine department.

Bottlings

BB	Bordeaux bottled
BE	Belgian bottled
CB	Château bottled
DB	Domain bottled
DRC	Domaine de la Romanée Conti
EA	Erzeugerabfüllung (Estate bottled German Wine)
ES	Estate bottled
NB	Dutch bottled
OB	Bottled in country of origin
SB	Swiss bottled
UK	Bottled in United Kingdom
XX	Bottling not known

Packing

(oc)	original carton
(owc)	original wooden case
(wc)	wooden case
(sc)	Sotheby's carton

General

[]	believed e.g. [1970] believed 1970
cm.	centimetres
doz.	dozen

Reprinted with permission from *Sotheby's,* London, England. Chart reprinted from the "Wednesday 18th November 1992" *Sotheby's,* London, catalogue, p. 5.9/91.

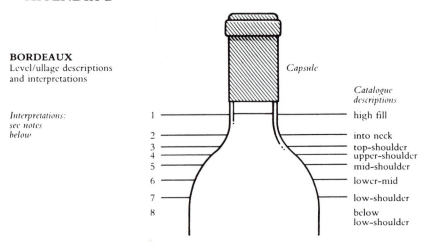

BORDEAUX
Level/ullage descriptions
and interpretations

Capsule

Catalogue descriptions

*Interpretations:
see notes
below*

1 — high fill

2 — into neck
3 — top-shoulder
4 — upper-shoulder
5 — mid-shoulder

6 — lower-mid

7 — low-shoulder

8 — below low-shoulder

1 *high fill:* normal fill. Level of young wines. Exceptionally good in wines over 10 years old.

2 *into neck:* can be level of fill. Perfectly good for any age of wine. Outstandingly good for a wine of 10 years in bottle, or longer.

3 *top-shoulder:* normal level for any claret 15 years old or older.

4 *upper-shoulder:* slight natural reduction through easing of cork and evaporation through cork and capsule. Usually no problem. Acceptable for any wine over 20 years old. Exceptional for pre-1940 wines.

5 *mid-shoulder:* probably some weakening of the cork and some risk. Not abnormal for wines 30/40 years of age. Estimates usually take this into account.

6 *lower-mid-shoulder:* some risk. Low estimates, usually no reserve.

7 *low-shoulder:* risky and usually only accepted for sale if wine or label exceptionally rare or interesting. Always offered without reserve and with low estimate.

8 *below low-shoulder:* not acceptable for sale unless a rare sort of bottle. Wine will usually be undrinkable.

BURGUNDY AND HOCK

Because of the slope of shoulder it is impractical to describe levels as mid-shoulder, etc. Wherever appropriate the level between cork and wine will be measured and catalogued in centimetres.

The condition and drinkability of burgundy is less affected by ullage than its equivalent from Bordeaux. For example, a 5 to 7 cm. ullage in a 50-year-old burgundy can be considered normal, indeed good for age, 3.5 to 4 cm. excellent for age, even 7 cm. rarely a risk.

CAUTION

Though every effort is made to describe or measure the levels of older vintages, corks over 20 years old begin to lose their elasticity and levels *can* change between cataloguing and sale. Old corks have also been known to fail during or after shipment.

We therefore repeat that there is always a risk of cork failure with old wines and due allowance must be made for this.

Important note: it is not Christie's policy to open original cases.

Under no circumstances can an adjustment of price or credit be made after delivery. (See also condition of sale A8).

Reprinted with permission from *Christie's,* London, England. Chart reprinted from *Christie's,* London, "Thursday 9 April 1992 Fine Wines and Vintage Port" catalogue, p. 2.

Index